PHYSICS IS LOGIC Part II:

The Theory of Everything, The Megaverse Theory of Everything, $U(4)\otimes U(4)$ Grand Unified Theory (GUT), Inertial Mass = Gravitational Mass, Unified Extended Standard Model and a New Complex General Relativity with Higgs Particles, Generation Group Higgs Particles

STEPHEN BLAHA

BLAHA RESEARCH

ISBN: 978-0-9893826-8-7

Cover Credits

The cover background is a copy of the painting "Horse Guards on parade" from the author's collection.

Rev. 00/00/01 October 15, 2015

To My Wife Margaret

Some Other Books by Stephen Blaha

All the Megaverse! Starships Exploring the Endless Universes of the Cosmos using the Baryonic Force (Blaha Research, Auburn, NH, 2014)

All the Universe! Faster Than Light Tachyon Quark Starships & Particle Accelerators with the LHC as a Prototype Starship Drive Scientific Edition (Pingree-Hill Publishing, Auburn, NH, 2011).

From Asynchronous Logic to The Standard Model to Superflight to the Stars (Blaha Research, Auburn, NH, 2011)

From Asynchronous Logic to The Standard Model to Superflight to the Stars; Volume 2: Superluminal CP and CPT, U(4) Complex General Relativity and The Standard Model, Complex Vierbein General Relativity, Kinetic Theory, Thermodynamics (Blaha Research, Auburn, NH, 2012)

The Algebra of Thought & Reality: The Mathematical Basis for Plato's Theory of Ideas, and Reality Extended to Include A Priori Observers and Space-Time; Second Edition (Pingree-Hill Publishing, Auburn, NH, 2009)

Quantum Big Bang Cosmology: Complex Space-time General Relativity, Quantum Coordinates, Dodecahedral Universe, Inflation, and New Spin 0, ½, 1 & 2 Tachyons & Imagyons™ (Pingree-Hill Publishing, Auburn, NH, 2004)

SuperCivilizations: Civilizations as Superorganisms (McMann-Fisher Publishing, Auburn, NH, 2010)

Standard Model Symmetries, And Four and Sixteen Dimension Complex Relativity; The Origin Of Higgs Mass Terms (Blaha Research, Auburn, NH, 2012)

The Bridge to Dark Matter; A New Sister Universe; Dark Energy; Inflatons; Quantum Big Bang; Superluminal Physics; An Extended Standard Model Based on Geometry (Blaha Research, Auburn, NH, 2013)

Universes and Megaverses: From a New Standard Model to a Physical Megaverse; The Big Bang; Our Sister Universe's Wormhole; Origin of the Cosmological Constant, Spatial Asymmetry of the Universe, and its Web of Galaxies; A Baryonic Field between Universes and Particles; Flatverse Extended Wheeler-DeWitt Equation (Blaha Research, Auburn, NH, 2014)

Available on bn.com, Amazon.com, Amazon.co.uk and other international web sites as well as at better bookstores (through Ingram Distributors).

Preface

This book is the second volume of *Physics is Logic*. It extends the discussion of volume I on the Generation group - the origin of the four fermion generations of each fermion species, and introduces new Higgs particles associated with the Generation group.

It develops a new formulation of Complex General Relativity, which yields both Einstein's General Relativity and a set of new Higgs particles that contribute to the masses of fermions. One result of this new formulation is that it can be united with The Extended Standard Model to form a Theory of Everything. It also proves, within the framework of this Theory of Everything that inertial mass equals gravitational mass – a concept that has been a question of interest since Newton's time, and was much considered by Einstein and colleagues.

The features of a analytically continued Complex General Relativity are also discussed in some detail.

In addition the Extended Standard Model previously developed by the author is shown to be a Grand Unified Theory (GUT) if extended to ultra high energies. Its GUT symmetry is shown to be $U(4) \otimes U(4)$.

Finally the author's proposed Megaverse of universes is discussed and the outlines of a new Megaverse Standard Model is developed. The interface between our universe and the Megaverse is described. Some points of interest include the proximity of every point of our universe to the Megaverse, the possibility of entry/exit to the Megaverse, and the possibility of quantum communication between points in our universe and points in the Megaverse.

CONTENTS

FIGURES

1. Short Summary of Blaha's Extended Standard Model Derivation

This chapter *briefly* summarizes the derivation of the Extended Standard Model presented in volume I. The Extended Standard Model follows ultimately from Asynchronous Logic, which leads to complex 4-dimensional space-time, which in turn leads directly to the form of the known Standard Model with extensions for Dark matter.

1.1 The Orgin of 4-dimensional Complex Space-Time

One of the basic requirements for a fundamental theory of elementary particles is the need to support parallel processes that are spatially separated. We see this when we consider complex Feynman diagrams; we see this in penomena on larger scales such as in atomic physics. This requirement requires a Logic that has four truth values[1] – Asynchronous Logic – which is used in designing extended, parallel computations, and computer architectures.

Expressing the four truth values as a 4-component spinor (and using a matrix notation for logic[2]) we find that a four-dimensional space-time is required. The space-time may have real or complex valued coordinates. With *qubits* in mind, and realizing the rich structure of The Standard Model, we assume complex space-time and thus the Complex Lorentz group.

Since our measurements of time and space always yield real values we introduce a local group of transformations, called the Reality group, that maps complex coordinates to real-valued coordinates. The Reality group must have 16 generators for complex 4-dimensional space-time. An examination of the U(4) group shows it has the subgroups SU(3), SU(2), U(1), another SU(2) and another U(1). The generators of these subgroups do not commute in general. The 16 generators of the product R = SU(3)⊗SU(2)⊗U(1)⊗SU(2)⊗U(1) algebraically span the set of generators of U(4) and so R can be chosen to be the Reality group generators.

1.2 Four Species of Fermions

The construction of the fermion fields begins with a four component spinor representing a truth value. Landauer and others have pointed out that a logic value has a minimum energy that we can take to be the mass associated with a spinor. This leads us to call a spinor with mass an *iota* particle, and to view the iota as the core, in some sense, of a fermion. Then defining a Dirac equation for a free fermion field at rest:

$$(\gamma^0 m - m)\psi = 0 \tag{1.1}$$

[1] The four truth values are true, false, intermediate (indefinte status but with some data present), and NULL (staus indefinite with no data present). See Fant (2005) for details. It is possible to use a two truth value logic supplemented by a synchronized clock. This formulation would be analogous to a 2-spinor formulation of the Dirac equation.

[2] Blaha (2010a).

we consider the possible states that can be obtained by *Complex Lorentz group boosts*. We find that there are exactly four types of states that can be obtained by boosts that meet the essential criterion for a fundamental, free, stable particle interpretation: the energy of the boosted state must be real-valued.[3] We call these four types of particles fermion *species*. Their characteristics and the known types of fermions, with which we identify them, are:

Type of Dirac-like Equation	**Standard Model Fermion Species**
Leptons	
Dirac equation with real energy and 3-momentum	Charged Leptons
Tachyon equation with real energy and 3-momentum ($m^2 < 0$)	Neutral Leptons (Neutrinos)
Quarks	
Dirac-like equation with real energy and complex 3-momentum	Up-type Quarks
Tachyon equation with real energy and complex 3-momentum ($m^2 < 0$)	Down-type Quarks

This classification of fermions into four species based on the Complex Lorentz group is a unique new feature that is susceptible to experiment and simply explains the hitherto inexplicable gross structure of the fermion spectrum.

A major benefit of this development is the parity violation inherent in the free tachyon equations. The parity violation appears in the dynamic equations of The Standard Model in a manner corrsponding to the known interactions of The Standard Model. Thus a fundamental justification for parity violation flows immediately from the application of Complex Lorentz group boosts to a free spinor equation for a particle at rest.

1.3 EletroWeak Doublets

The Complex Lorentz group can be used to establish a doublet structure that later is seen to become the ElectroWeak doublets. First, we note that Complex Lorentz group transformations can change a Dirac type particle into a tachyonic particle and *vice versa*. Thus we can group Dirac fermions and tachyon fermions into doublets for both leptons and quarks (separately). If one then groups a Dirac fermion and a tachyon fermion into a doublet state, the dynamic equations will admit of this rotation and transform correctly. However if one forms a free lagrangian for these equations in a 2×2 matrix formulation, then one sees that the lagrangian is *not* invariant under this transformation.[4] Thus the species cannot be transformed into one another. Consequently we have a doublet structure for leptons. We also have a doublet structure for quarks. But, we emphasize, transformations between particles and tachyons is not a symmetry – precluding this possible symmetry in the eventual Standard Model.

---

[3] If the energy were not real-valued the particle would be able to decay – contrary to the assumption that the particle is fundamental.

[4] See pp.228-231 of Blaha (2011c) and volume I.

1.4 SU(2)⊗U(1) Symmetries Due to Real and Complex Valued Velocity Boosts

Complex Lorentz *boosts* using real-valued velocities (both subluminal and superluminal) lead to ElectroWeak SU(2)⊗U(1). Complex Lorentz boosts using complex-valued velocities (both subluminal and superluminal) lead to Dark ElectroWeak symmetry using the other SU(2)⊗U(1) symmetry in the Reality group R. We associate the other SU(2)⊗U(1) symmetry with Dark matter and so call it the Dark symmetry. This symmetry leads to Dark particles, Dark ElectroWeak doublets and interactions.

The source of these symmetries is in the Reality group transformations required to make the complex coordinates of boosted fermions real-valued. The symmetries of the factors of R have local Yang-Mills fields from which emerges the ElectroWeak sector of The Extended Standard Model. We call it "extended" since it incorporates a new Dark ElectroWeak sector (that remains to be experimentally verified.)

1.5 Quarks have Complex-Valued Velocities that Lead to Color SU(3)

The 3-momenta of up-type and down-type quarks are complex valued and consist of orthogonal real and imaginary components. A study of the transformations of these complex velocities shows that they have an SU(3) symmetry due to their complex-valued nature. This SU(3) symmetry is masked by SU(3) transformations of the Reality group. But it is nevertheless present and should be experimentally found – perhaps in LHC studies of quark-gluon plasmas created in high energy ion-ion collisions, or perhaps in a re-analysis of deep inelastic electron-nucleon parton models. The SU(3) symmetry is realized by the strong interaction SU(3).

The method of our derivation shows why quarks – and not leptons – have the SU(3) strong interaction. Leptons have purely real-valued velocities.

It appears that Dark quarks are SU(3) singlets since no Dark-normal particle bound states have been found.

At this point we have four species of normal fermions and four similar species of Dark fermions in a one generation Standard Model. We now introduce generations.

1.6 Four Fermion Generations

Up to this point the Standard Model properties that we have found would apply to one generation, or multi-generation, versions of The Standard Model. We now turn to the fermion generation question.

We begin by noting the longstanding questions of whether baryon number and lepton number are (separately) conserved. Experimentally they appear conserved. However there is theoretical concern because, unlike charge and electromagetism, baryonic and leptonic forces have not apparently been found. We say apparently because major discrepancies in measurements of the gravitatiional constant G have been found. The discrepancies are large enough to justify a claim that an ultra weak baryonic force exists with a coupling constant of the order of Gm_H^2 where m_H is the mass of a hydrogen atom. Assuming that baryonic, leptonic,

Dark baryonic and Dark leptonic forces exist with corresponding gauge fields,[5] and number operators; B, L, B_D, L_D; we see that a U(4) symmetry also exists since linear combinations these four number operators would also be conserved:

$$\begin{aligned} B' &= aB + bL + cB_D + dL_D \\ L' &= eB + fL + gB_D + hL_D \\ B_D' &= iB + jL + kB_D + lL_D \\ L_D' &= mB + nL + oB_D + pL_D \end{aligned} \tag{1.2}$$

The primed number operators are conserved if the unprimed numbers are conserved.

This U(4) symmetry would also apply to particle fields and particles. We have taken the four normal particle species to each be singlets under this new symmetry. We now assume each species is not one particle but rather four particles in a 4-dimensional U(4) representation. Since the four particles of each species have very different masses, the U(4) symmetry is badly broken.

Thus we now have a four generation Extended Standard Model, except for the Higgs sector, which we will address next.

1.7 Higgs Sector

The Higgs sector of The Extended Standard Model is similar to that of the known Standard Model except that it adds more terms to the fermion mass matrices for each of the eight fermion species (four normal matter species and four Dark matter species). Chapter 5 of this volume lists the total mass matrices of each species. These matrices must be diagonalized to find the physical species generations' masses. Quoting a section in chapter 5 below:

"Combining the terms in eq. 5.56 for each species we obtain their total mass matrices below which can then be diagonalized to obtain the masses of the fermions within each species.

Charged Lepton Species Total Mass Matrix

$$m_{etot} = m_{EWe} + m_{Ge} \tag{5.57}$$

Neutral Lepton Species Mass Matrix

$$m_{\upsilon tot} = m_{EW\upsilon} + m_{G\upsilon} \tag{5.58}$$

Up-Type Quark Species Mass Matrix

$$m_{utot} = m_{EWu} + m_{Uu} + m_{Gu} \tag{5.59}$$

Down-Type Quark Species Mass Matrix

$$m_{dtot} = m_{EWd} + m_{Ud} + m_{Gd} \tag{5.60}$$

[5] Due to symmetry breaking the baryonic fields are found to be massless, long range fields; the leptonic fields are found to be massive, short range fields in volume I.

Dark Charged Lepton Species Total Mass Matrix

$$m_{Detot} = m_{DEWe} + m_{DGe} \qquad (5.61)$$

Dark Neutral Lepton Species Mass Matrix

$$m_{D\upsilon tot} = m_{DEW\upsilon} + m_{DG\upsilon} \qquad (5.62)$$

Dark Up-Type Quark Species Mass Matrix

$$m_{Dutot} = m_{DEWu} + m_{DUu} + m_{DGu} \qquad (5.63)$$

Dark Down-Type Quark Species Mass Matrix

$$m_{Ddtot} = m_{DEWd} + m_{DUd} + m_{DGd} \qquad (5.64)$$

We now note that the preceding formal development yields $m_{Ge} = m_{G\upsilon} = m_{Gu} = m_{Gd} = m_{DGe} = m_{DG\upsilon} = m_{DG\upsilon} = m_{DGu} = m_{DGd} = m_G$. The gravitational contribution to all fermions of all all species is the same.

Moreover, the gravitational contribution to each fermion mass sets the scale for all fermion masses (and secondarily of massive gauge bosons' masses) yielding the "principle" of Newton, Einstein and others that *inertial mass equals to gravitational mass.* (See chapter 6.)

The generation group contributions, in the spontaneous breakdown that we described, appear only in quark and Dark quark mass matrices providing, possibly, a reason why quark masses are so much larger than lepton masses.

The mass matrices above can each be diagonalized in a manner similar to that of eqs. 16.50 and 16.51 in chapter 2 and volume I."

The mass matrices above have the usual ElectroWeak mass matrices.

Mass matrix contributions from the generation group breakdown are m_{Uu}, m_{Ud}, m_{DUu}, and m_{DUd}. These mass matrices may be the reason why quark masses are much greater than lepton masses.

Mass matrix contributions from Complex General Relativity, which has a form supporting new Higgs scalar bosons, appear in all species mass matrices. They have the form m_{DGi} where i specifies the species. In chapters 5 and 6 we suggest they resolve the question of the equality of inertial and gravitational mass. This longstanding question ranging back to Newton has been a subject of much discusion. The appearance of a mass contribution from Complex General Relativity in all species mass matrices sets the mass scale of elementary particles since all the other terms in each total mass matrix must have the same scale.

1.7 Extended Standard Model

At this point we have derived The Extended Standard Model which includes the known Standard Model and adds a Dark Matter sector. It also explains the equality of inertial and gravitational mass setting the stage for the Theory of Everything derived later in this volume.

1.8 Perturbation Theory

Having derived the form of The Extended Standard Model we now address the issue of infinities that appear in perturbation theory calculations. These high energy infinities are eliminated by t'Hooft's renormalization program in the ElectroWeak sector. However the well-known triangle anomaly persists, and any extension to include General Relativity in a unified theory still has its well-known infinities to resolve. Further we intend to consider extensions of the theory to 16-dimensional space-time. For all these reasons it appears necessary to use the Two-Tier quantum field theory formulation which eliminates all infinities in the Extended Standard Model, Quantum Gravity, and in higher dimensions. Two-Tier QFT gives the well-known results of lw energy QED calculations and eliminates infinities at high energy Feynman diagram by Feynman diagram.

An added advantage, shown in volume I, is that Two-Tier gauge particles could fuel the expansion of the universe from the Big Bang point including the inflationary phase. Just as they eliminate field theory infinities, they also eliminate the Big Bang infinities at t = 0.

The naturalness of our derivation, and its rapid explication of Standard Model features, exemplifies R. P. Feynman's dictum: *The correct answer is usually self-evident **after** it is found.*

2. The Generation Group, Fermion Generations and New Higgs Fields

In volume I we established that complex Lorentz group boosts imply four species of normal fermions and four species of Dark fermions in the absence of interactions due to the requirement that fundamental fermions must have real-valued energies. (Complex-valued energies would imply that fermions decay into "something else" contrary to the assumption that they are fundamental.) Subsequently we developed a one generation Standard Model.

We then noted that the one generation theory and, more importantly, experiments indicate that baryon number, and lepton number, was conserved to great accuracy. Extending these conservation laws to Dark Matter we assume that Dark baryon number and Dark lepton number are also conserved. Baryon number is the sum of the up and down quark baryon numbers in a state of normal matter. Similarly Dark matter baryon number is the sum of the up and down Dark quark baryon numbers in a Dark state. Lepton number is the sum of the charged and neutral leptons in a state of normal matter as is Dark lepton number for a Dark matter state. As a result there is a quadruple of conserved number laws. Any linear combination of this quadruple is also conserved:

$$\begin{aligned}
B' &= aB + bL + cB_D + dL_D \\
L' &= eB + fL + gB_D + hL_D \\
B_D' &= iB + jL + kB_D + lL_D \\
L_D' &= mB + nL + oB_D + pL_D
\end{aligned} \tag{2.1}$$

or

$$N' = AN \tag{2.2}$$

where the coefficients can be real or complex. Thus there is a (broken) U(4) transformation group of particle number operators since we require inverse transformations to be the hermitean conjugates of transformations. Further this transformation group can be made local in the sense of Yang and Mills so that a local U(4) symmetry group emerges.

The Yang-Mills *ansatz* for local particle symmetries is appropriate in this case as well since the distinction between the four particle number types can be viewed as locally determined.

The fermions of the one generation Standard Model are U(4) scalars. Generalizing to the multi-generation Standard Model, we assume the particles of each species to be U(4) vectors – which is a reasonable assumption in accord with Leibniz's Minimax Principle since this assumption opens the possibility of varying masses and transformations between different generations via interactions. Then we find four generations of each fermion and Dark fermion species.

Three generations of each normal fermion species have been found experimentally. There appear to be experimental discrepancies in new data that are possibly experimental hints of a fourth lepton generation.[6]

This U(4) symmetry group, which we call the Generation group, is badly broken: 1) it is broken between normal and Dark baryons since normal baryons (quarks) are color SU(3) triplets and Dark quarks are color SU(3) singlets; 2) the masses of the various generations of each species differ due to the Higgs Mechanism.

The U(4) Generation group leads to 16 Yang-Mills gauge fields that embody the Generation group interactions. See chapter 16 of volume I for a discussion of Generation group symmetry breaking via the Higgs Mechanism and the resulting long range and short interactions that result. The Baryon Number field and the Dark Baryon Number field are massless long range fields that play a role in the dynamics of our universe and possibly between universes in the Megaverse. See volume I for details.

2.1 Strength of the Baryonic and Dark Baryonic Interaction

Volume I contains an estimate of the strength of the baryonic coulomb force obtained by an analysis of the substantial discrepancies between various measurements of the gravitational constant G. Attributing the discrepancies to the baryonic force, due to varying baryon number in measurement weights, we found the baryonic coulomb coupling constant to approximately be

$$\alpha_B = \beta^2/4\pi \simeq 0.118\ Gm_H{}^2 \qquad (2.3)$$

where the baryonic coulomb force between matter samples with n_1 and n_2 baryons respectively is

$$F_B = (\beta^2/4\pi)n_1n_2/r^2 \qquad (2.4)$$

and where m_H is the mass of a hydrogen atom. Eq. 2.3 indicates a very weak baryonic force much smaller than the gravitational force. The baryon "fine structure constant" is miniscule in comparison to the electromagnetic fine structure constant $\alpha \simeq 1/137$.

The indicated smallness of the baryonic force and thus of the Generation group coupling constant has a significant impact on the possibility of a GUT (Grand Unified Theory) for the Extended Standard Model. While one can visualise a unification energy M_{GUT} of the order of 10^{16} GeV to 10^{17} GeV for the SU(3)⊗SU(2)⊗U(1)⊗SU(2)⊗U(1) interactions, the

---

6 R. Aaij et al, Phys. Rev. Lett. **115**, 111803 (2015); J. P. Lees et al, Phys. Rev. Lett. **109**, 101802 (2012).

value of the U(4) Generation group coupling constant makes its participation in a unification unlikely. In chapter 4 we therefore suggest the GUT unification leads to a U(4)⊗U(4) GUT.

2.2 Gauge Fields and Higgs Fields of the Generation Group

The generation group U(4) gauge fields and Higgs fields were discussed in chapter 16 of volume I. This subsection contains that chapter and adds some clarifying comments in *itallics.*

Enhanced Chapter 16 of Volume I
with abovementioned material added and italicized:

16.1.1 Two-Tier Lepton Sector *Normal and Dark Covariant Derivatives*

We begin with the definition of a quadruplet of leptons – a pair of doublets, one normal and one Dark, instead of a single doublet. We define left and right lepton quadruplets with[7]

$$\Psi_{L,Ra}(X) = \begin{bmatrix} \psi_{DL,Ra}(X) \\ \psi_{NL,Ra}(X) \end{bmatrix} \tag{16.1}$$

where a is a generation index ranging from 1 to 4, $\psi_{NL,R}(X)$ is a "normal" ElectroWeak-like lepton doublet, and where $\psi_{DL,R}(X)$ is a Dark ElectroWeak-like lepton doublet consisting of a Dark electron-like fermion and a Dark neutrino-like fermion.

We define covariant derivative terms which we express in matrix form are

$$D_{L,R}(X) = \begin{bmatrix} \gamma^\mu D_{DL,R\mu} & 0 \\ 0 & \gamma^\mu D_{NL,R\mu} \end{bmatrix} \tag{16.2}$$

where the normal matter left-handed covariant derivative is

$$D_{NL\mu} = \partial/\partial X^\mu - \tfrac{1}{2} ig\boldsymbol{\sigma}\cdot\mathbf{W}_\mu + \tfrac{1}{2} ig' B_\mu - \tfrac{1}{2} ig_G \mathbf{G}\cdot\mathbf{U}_\mu \tag{16.3}$$

where g_G is an ultra-weak generational coupling constant, $\mathbf{G}\cdot\mathbf{U}_\mu$ is the sum of the inner product of 16 U(4) generators G_i and U(4) gauge fields $U_i(X)$, and where the Dark matter left-handed covariant derivative is

---

[7] The X's are Two-Tier coordinates.

$$D_{DL\mu} = \partial/\partial X^{\mu} - \tfrac{1}{2} i g_D \boldsymbol{\sigma}\cdot\mathbf{W}'_{\mu} + \tfrac{1}{2} i g_D' B'_{\mu} + \tfrac{1}{2} i g_D'' B_{\mu} - \tfrac{1}{2} i g_G \mathbf{G}\cdot\mathbf{U}_{\mu} \qquad (16.4)$$

with $\boldsymbol{\sigma}$ a vector composed of the Pauli matrices. The right-handed covariant derivatives have a simpler form. The normal matter right-handed covariant derivative is

$$D_{NR\mu} = \partial/\partial X^{\mu} + \tfrac{1}{2} i g' B_{\mu} - \tfrac{1}{2} i g_G \mathbf{G}\cdot\mathbf{U}_{\mu} \qquad (16.5)$$

and the Dark matter right-handed covariant derivative is

$$D_{DR\mu} = \partial/\partial X^{\mu} + \tfrac{1}{2} i g_D' B'_{\mu} + \tfrac{1}{2} i g_D'' B_{\mu} - \tfrac{1}{2} i g_G \mathbf{G}\cdot\mathbf{U}_{\mu} \qquad (16.6)$$

The normal and Dark electroweak fields above are functions of a Two-Tier X. The Faddeev-Popov mechanism operative for these types of fields is described in appendix 19-A, and in chapter 12, of Blaha (2011c).

The modified covariant derivatives should be inserted in the leptonic sector lagrangian terms of the Extended Standard Model presented in volume I.

16.1.2 Quark Sector

In the *quark* sector we define left and right quark quadruplets with

$$\Psi_{qL,Ra}(X_c) = \begin{bmatrix} \psi_{DqL,Ra}(X_c) \\ \psi_{NqL,Ra}(X_c) \end{bmatrix} \qquad (16..7)$$

where $\psi_{NqL,Ra}(X_c)$ is a "normal" ElectroWeak-like quark doublet, and where $\psi_{DqL,Ra}(X_c)$ is a Dark ElectroWeak-like quark doublet consisting of a SU(3) singlet Dark up-quark of unit Dark charge and a SU(3) singlet Dark down-quark of zero Dark charge in the ath generation.

The covariant derivative terms are contained in $D_q(X_c)$ which we express in matrix form as

$$D_{qL,R}(X_c) = \begin{bmatrix} \gamma^{\mu} D_{qDL,R\mu}(X_c) & 0 \\ 0 & \gamma^{\mu} D_{qNL,R\mu}(X_c) \end{bmatrix} \qquad (16.8)$$

where the normal quark matter left-handed covariant derivative is

$$D_{qNL\mu} = \partial/\partial X_c^{\mu} - \tfrac{1}{2} i g \boldsymbol{\sigma}\cdot\mathbf{W}_{\mu} - i g' B_{\mu}/6 - \tfrac{1}{2} i g_G \mathbf{G}\cdot\mathbf{U}_{\mu} + i g_C \boldsymbol{\tau}\cdot A_{C\mu} \qquad (16.9)$$

and where the Dark quark left-handed covariant derivative is

$$D_{qDL\mu} = \partial/\partial X_c{}^\mu - \tfrac{1}{2}ig_D\boldsymbol{\sigma}\cdot\mathbf{W'}_\mu + \tfrac{1}{2}ig_D'B'_\mu + \tfrac{1}{2}ig_D''B_\mu - \tfrac{1}{2}ig_G\mathbf{G}\cdot\mathbf{U}_\mu \qquad (16.10)$$

since Dark quarks are SU(3) singlets with unit or zero Dark charge. The right-handed quark covariant derivatives have a simpler form. The normal quark right-handed covariant derivative is

$$D_{qNR\mu} = \partial/\partial X_c{}^\mu + \tfrac{1}{2}ig'B_\mu/3 - \tfrac{1}{2}ig_G\mathbf{G}\cdot\mathbf{U}_\mu + ig_C\tau\cdot\mathbf{A}_{C\mu} \qquad (16.11)$$

and the Dark quark right-handed covariant derivative is

$$D_{qDR\mu} = \partial/\partial X_c{}^\mu + \tfrac{1}{2}ig_D'B'_\mu + \tfrac{1}{2}ig_D''B_\mu - \tfrac{1}{2}ig_G\mathbf{G}\cdot\mathbf{U}_\mu \qquad (16.12)$$

The normal and Dark gauge boson fields are functions of $X_c. = (X_{r_\mu}(y_r), X_{i_\mu}(y_i))$ of eqs. 14.11 and 14.12. The Faddeev-Popov mechanism is operative for gauge boson fields and is described in appendix 19-A of Blaha (2011c).[8] The complexon quark Extended Standard Model ElectroWeak Sector covariant derivatives in quadruplet matrix form are

$$D_{qL,R}(X_c) = \begin{bmatrix} \gamma^\mu D_{qDL,R\mu} & 0 \\ 0 & \gamma^\mu D_{qNL,R\mu} \end{bmatrix} \qquad (16.13)$$

The remaining parts of the complexon Standard Model are described in chapter 23 of Blaha (2011) and summarized below. The addition of singlet Dark quark Higgs terms is also required.

The lagrangian density and action is

$$\mathcal{L}_{CSM} = \Psi_L{}^a{}^\dagger\gamma^0 i\gamma^\mu D_{L\mu}\Psi_L{}^a - \Psi_R{}^a{}^\dagger\gamma^0 i\gamma^\mu D_{R\mu}\Psi_{3R}{}^a + \Psi_{CL}{}^a{}^\dagger\gamma^0 i\gamma^\mu \mathcal{D}_{qL\mu}\Psi_{CL}{}^a + \Psi_{CR}{}^a{}^\dagger\gamma^0 i\gamma^\mu \mathcal{D}_{qR\mu}\Psi_{CR}{}^a -$$
$$- \mathcal{L}_{BareMasses} + \mathcal{L}_{Gauge} + \mathcal{L}_{Mass} + \mathcal{L}_{Ufields} \qquad (16.14)$$

where a is the generation index. $\mathcal{L}_{BareMasses}$ contains the fermion bare mass terms. Also,

$$\mathcal{L}_{Gauge} = \mathcal{L}_{GaugeEW} + \mathcal{L}_{GaugeC} + \mathcal{L}_{GaugeEWD} \qquad (16.15)$$

with

$$\mathcal{L}_{GaugeEW} = -\tfrac{1}{4} F_W{}^{a\mu\nu}F_W{}^a{}_{\mu\nu} - \tfrac{1}{4} F_B{}^{\mu\nu}F_{B\mu\nu} + \mathcal{L}_{EW}{}^{ghost} \qquad (16.16)$$

$$\mathcal{L}_{GaugeEWD} = -\tfrac{1}{4} F'_W{}^{a\mu\nu}F'_W{}^a{}_{\mu\nu} - \tfrac{1}{4} F_{B'}{}^{\mu\nu}F_{B'\mu\nu} + \mathcal{L}_W{}^{ghost} \qquad (16.17)$$

---

[8] Those who might be concerned about the propagator term $<W_i(X), W_j(X_c)>$ and similar propagators where one field is a function of X and the other field is a function of X_c should note that such terms are to very good approximation equal to $<W_i(X), W_j(X)>$ for energies much less than M_c (which could be as large as the Planck energy.)

and

$$\mathcal{L}_{GaugeC} = \mathcal{L}_{CCG} + \mathcal{L}_{C}^{ghost} + \mathcal{L}_{CC}^{ghost} \tag{16.18}$$

$$\mathcal{L}_{Ufields} = -\tfrac{1}{4} F_U{}^{a\mu\nu} F_{U\mu\nu} + \mathcal{L}_U^{ghost} + \mathcal{L}_U^{UHiggs} \tag{16.19}$$

where $\mathcal{L}_U^{UHiggs}$ is discussed in section 16.4. The ElectroWeak gauge bosons W_μ^a, B_μ and B'_μ field tensors are:

$$F_W{}^a{}_{\mu\nu} = \partial W^a{}_\mu/\partial X^\nu - \partial W^a{}_\nu/\partial X^\mu + g_2 f^{abc} W^b{}_\mu W^c{}_\nu \tag{16.20}$$

$$F_{B\mu\nu} = \partial B_\mu/\partial X^\nu - \partial B_\nu/\partial X^\mu \tag{16.21}$$

and the Dark ElectroWeak gauge bosons W'^a_μ and B'_μ field tensors are:

$$F_{B'\mu\nu} = \partial B'_\mu/\partial X^\nu - \partial B'_\nu/\partial X^\mu$$

$$F'_W{}^a{}_{\mu\nu} = \partial W'^a{}_\mu/\partial X^\nu - \partial W'^a{}_\nu/\partial X^\mu + g_2 f^{abc} W'^b{}_\mu W'^c{}_\nu \tag{16.22}$$

The U fields tensors are:

$$F_U{}^a{}_{\mu\nu} = \partial U^a{}_\mu/\partial X^\nu - \partial U^a{}_\nu/\partial X^\mu + g_G f_4{}^{abc} U^b{}_\mu U^c{}_\nu \tag{16.23}$$

where $f_4{}^{abc}$ are the U(4) algebra commutator constants.

$\mathcal{L}_{EW}^{ghost}$ contains the Faddeev-Popov ghost terms for the ElectroWeak W_μ^a gauge bosons. The complexon color gluon lagrangian $\mathcal{L}_{CCG}$ is defined by

$$\mathcal{L}_{CCG} = -\tfrac{1}{4} F_{CC}{}^{a\mu\nu}(X) F_{CC}{}^a{}_{\mu\nu}(X) \tag{16.24}$$

where

$$F_{CC}{}^a{}_{\mu\nu} = \partial/\partial X_c{}^\nu A_C{}^a{}_\mu - \partial/\partial X_c{}^\mu A_C{}^a{}_\nu + g f_{su(3)}{}^{abc} A_C{}^b{}_\mu A_C{}^c{}_\nu \tag{16.25}$$

where $A_C{}^a{}_\nu$ is the color gluon gauge field, g is the color coupling constant, and the $f_{su(3)}{}^{abc}$ are the SU(3) structure constants.

In addition $\mathcal{L}_C^{ghost}$ is the color SU(3) Faddeev-Popov ghost terms defined in appendix 19-A of Blaha (2011c) for the complexon Lorentz gauge and $\mathcal{L}_{CC}^{ghost}$ is the complexon color SU(3) constraint ghost terms defined through the Faddeev-Popov mechanism. The mass sector $\mathcal{L}_{Mass}$ is presumably based on the Higgs Mechanism.which creates the *fermion masses, the* ElectroWeak vector boson masses, and generation mixing.

The lagrangian is supplemented with the following condition on all complexon fields $\Phi_{\ldots}$:[9]

[9] These conditions implement the orthogonality of the real and imaginary parts of complexon 3-momentum.

$$\nabla_r \cdot \nabla_i \Phi\ldots = 0 \tag{16.26}$$

Non-complexon fields Ω... in the left-handed formulation under consideration satisfy the subsidiary condition:

$$[\nabla_r \cdot \nabla_i - (\nabla_r^2 \nabla_i^2)^{1/2}]\Omega\ldots = 0 \tag{16.27}$$

which guarantees a complexon's real momentum is parallel to its imaginary momentum.

At first glance it would appear that the U field gauge terms appearing in covariant derivatives above would cause transitions between normal and Dark matter. This is true. However the extreme weakness of the U-field coupling constant, and thus its interactions, (much below gravity) precludes this interaction from being of any significance for the foreseeable future.

16.2 Generation U(4) Gauge Symmetry Breaking and Long Range Forces

In chapter 15 we showed that there was good experimental evidence for a conserved Baryon Number B and we proceeded to develop a simple U(1) gauge theory that would imply Baryon Number conservation in a manner analogous to QED's implying electric charge conservation. In section 16.1 we used a new symmetry group local U(4) to generalize the one generation Extended Standard Model to a four generation Extended Standard Model based on four conserved particle numbers: B, L, B_D, and L_D.[10]

We now assume in our construction that the four generation Extended Standard Model has a local U(4) symmetry that is broken by mass terms gewnerated by the Higgs Mechanism.

Further, we will assume that the Higgs breakdown yields two massless (long range) fields which we associate with Baryon Number B and Dark Baryon Number B_D. The remaining fields acquire masses and generate short range forces.

We use the following U(4) diagonal matrices:

$$\begin{aligned} G_1 &= \mathrm{diag}(1, 1, 1, 1) \\ G_2 &= \mathrm{diag}(0, 1, 0, 0) \\ G_3 &= \mathrm{diag}(0, 0, 1, 0) \\ G_4 &= \mathrm{diag}(0, 0, 0, 1) \end{aligned} \tag{16.28}$$

The U(4) algebra has 16 hermitean matrices that satisfy

$$G_i^\dagger = G_i \tag{16.29}$$

The particle numbers can be expressed in terms of the diagonal generators as

$$B = G_1 - G_2 - G_3 - G_4 \tag{16.30}$$

[10] Charge, although a conserved number, is a part of the ElectroWeak sector, account of which has already been taken.

$$B_D = G_2$$
$$L = G_3$$
$$L_D = G_4$$

The covariant derivatives have the general form *for both normal and Dark U gauge fields*:

$$D_{\ldots\mu} = \partial/\partial X^\mu + \ldots - \tfrac{1}{2} i g_G \mathbf{G}\cdot\mathbf{U}_\mu \tag{16.31}$$

where the ellipses indicate the other details of the particular covariant derivative. We now wish to express the four gauge fields $U_i(X)$ for $i = 1, 2, 3, 4$ corresponding to the diagonal generators in terms of the fields of the four particle number gauge fields: B_μ, L_μ, $B_{D\mu}$, and $L_{D\mu}$.

$$U_{i\mu} = A_{ik} N_{k\mu} \tag{16.32}$$

where A_{ik} are the elements of a matrix of constants and

$$N_\mu = \begin{bmatrix} B_\mu(X) \\ L_\mu(X) \\ B_{D\mu}(X) \\ L_{D\mu}(X) \end{bmatrix} \tag{16.33}$$

is a column vector consisting of the gauge fields corresponding to each of the conserved particle numbers.

The matrix A must have non-zero determinant so that eq. 16.32 can be inverted to express the particle number fields in terms of the four $U_i(X)$ gauge fields:

$$N_\mu = A^{-1} U_\mu \tag{16.34}$$

Resulting in

$$\begin{aligned} B_\mu(X) &= U_{1\mu} \\ L_\mu(X) &= U_{1\mu} + U_{2\mu} \\ B_{D\mu}(X) &= U_{1\mu} + U_{3\mu} \\ L_{D\mu}(X) &= U_{1\mu} + U_{4\mu} \end{aligned} \tag{16.35}$$

Then

$$D_{\ldots\mu} = \partial/\partial X^\mu + \ldots - \tfrac{1}{2} i g_G \Big[\sum_{i=5}^{16} \mathbf{G}_i \mathbf{U}_{i\mu} + B B_\mu(X) + L L_\mu(X) + B_D B_{D\mu}(X) + L_D L_{D\mu}(X)\Big] \tag{16.36}$$

where the particle numbers, which are analogous to the charges Q and Q' in ElectroWeak theory, are B, L, B_D, and L_D. They are expressed in terms of U(4) generators by eqs. 16.30.

16.3 Higgs Mass Mechanism for U(4) Generation Gauge Fields

We now require that there are two massless fields, one coupled to Baryon number and one coupled to Dark Baryon number. The Dark sector is assumed to be analogous to the normal particle sector in this respect. There are fourteen remaining fields that acquire masses and longitudinal components. These fields become short range, ultra-weak generational forces. The masses they acquire through the Higgs Mechanism are presumably very large as these gauge particles have not been found experimentally.[11]

We assume that a scalar Higgs field exists which is a U(4) vector with four components corresponding to the fermion generations. It is an SU(2)⊗U(1)⊗SU(3) ElectroWeak scalar. Its lagrangian density is

$$\mathcal{L}_U{}^{UHiggs} = (\partial\eta^\dagger/\partial X^\mu)(\partial\eta/\partial X^\mu) - \lambda(\eta^\dagger\eta - \rho^2)^2 + \mathcal{L}_U{}^{UHiggs}{}_{FermionMasses}$$

where $\mathcal{L}_U{}^{UHiggs}{}_{FermionMasses}$ are the fermion masses produced by the U Higgs Mechanism and where we choose a unitary gauge in which

$$\rho = \begin{bmatrix} 0 \\ \rho_1 \\ 0 \\ \rho_2 \end{bmatrix} \qquad (16.37)$$

where ρ_1 and ρ_2 are real fields. Then the covariant derivative of ρ is

$$D_{\ldots\mu}\,\rho = \{\partial/\partial X^\mu + \ldots - \tfrac{1}{2}ig_G[\Sigma\mathbf{G}_i\mathbf{U}_{i\mu} + BB_\mu(X) + LL_\mu(X) + B_DB_{D\mu}(X) + L_DL_{D\mu}(X)]\}\begin{bmatrix} 0 \\ \rho_1 \\ 0 \\ \rho_2 \end{bmatrix} \qquad (16.38)$$

The sum over i is from 5 through 16, and $[\mathbf{G}_i]_{jk}$ is the jk element of $\mathbf{G}_i$. Then

$$D_{\ldots\mu}\,\rho = \begin{bmatrix} -\tfrac{1}{2}ig_G\{\rho_1\Sigma[\mathbf{G}_i]_{12}\mathbf{U}_{i\mu} + \rho_2\Sigma[\mathbf{G}_i]_{14}\mathbf{U}_{i\mu}\} \\ \partial\rho_1/\partial X^\mu - \tfrac{1}{2}ig_G\rho_1L_\mu - \tfrac{1}{2}ig_G\{\rho_1\Sigma[\mathbf{G}_i]_{22}\mathbf{U}_{i\mu} + \rho_2\Sigma[\mathbf{G}_i]_{24}\mathbf{U}_{i\mu}\} \\ -\tfrac{1}{2}ig_G\{\rho_1\Sigma[\mathbf{G}_i]_{32}\mathbf{U}_{i\mu} + \rho_2\Sigma[\mathbf{G}_i]_{34}\mathbf{U}_{i\mu}\} \\ \partial\rho_2/\partial X^\mu - \tfrac{1}{2}ig_G\rho_2L_{D\mu} - \tfrac{1}{2}ig_G\{\rho_1\Sigma[\mathbf{G}_i]_{42}\mathbf{U}_{i\mu} + \rho_2\Sigma[\mathbf{G}_i]_{44}\mathbf{U}_{i\mu}\} \end{bmatrix} \qquad (16.39)$$

[11] Section 16.4 discusses this topic in more detail.

$$= \begin{bmatrix} -\tfrac{1}{2}ig_G\Sigma\{\rho_1[\mathbf{G}_i]_{12} + \rho_2[\mathbf{G}_i]_{14}\}\mathbf{U}_{i\mu} \\ \partial\rho_1/\partial X^\mu - \tfrac{1}{2}ig_G\rho_1 L_\mu - \tfrac{1}{2}ig_G\rho_2\Sigma[\mathbf{G}_i]_{24}\mathbf{U}_{i\mu} \\ -\tfrac{1}{2}ig_G\Sigma\{\rho_1[\mathbf{G}_i]_{32} + \rho_2[\mathbf{G}_i]_{34}\}\mathbf{U}_{i\mu} \\ \partial\rho_2/\partial X^\mu - \tfrac{1}{2}ig_G\rho_2 L_{D\mu} - \tfrac{1}{2}ig_G\rho_1\Sigma[\mathbf{G}_i]_{42}\mathbf{U}_{i\mu} \end{bmatrix} \tag{16.40}$$

since the generators $\mathbf{G}_i$ have zeroes along their diagonals for i = 5, … , 16.

From eq. 16.39 we find the corresponding Higgs field kinetic terms in the lagrangian are

$$(D_{\ldots\mu}\,\rho)^\dagger D_{\ldots}^{\;\mu}\,\rho = \partial\rho_1/\partial X^\mu\partial\rho_1/\partial X_\mu + \partial\rho_2/\partial X^\mu\,\partial\rho_2/\partial X_\mu + g_G^{\,2}\rho_1^{\,2}L_\mu\,L^\mu/4 + g_G^{\,2}\rho_2^{\,2}L_{D\mu}\,L_D^{\,\mu}/4 + \ldots \tag{16.41}$$

Note there are differing mass squared terms for the Lepton ($g_G^{\,2}\rho_1^{\,2}/4$) and Dark Lepton ($g_G^{\,2}\rho_2^{\,2}/4$) gauge fields making them short range fields with the likelihood of very large masses much beyond ElectroWeak gauge field masses, and with an ultra weak coupling constant g_G as suggested by the "experimental" coupling for the Baryonic force given in eq. 15.6.

The Baryonic and Dark Baryonic gauge fields are massless and thus long range although their coupling constant appears to be ultra weak – much below the gravitational coupling constant G.

We now turn to calculating the remaining terms in eq. 16.41 that determine the masses of the remaining 14 gauge fields. We begin by assigning matrix elements for the remaining hermitean U(4) generators:

$$\begin{aligned}
[G_5]_{ik} &= \delta_{i1}\delta_{k2} + \delta_{i2}\delta_{k1} \\
[G_6]_{ik} &= -i\delta_{i1}\delta_{k2} + i\delta_{i2}\delta_{k1} \\
[G_7]_{ik} &= \delta_{i1}\delta_{k3} + \delta_{i3}\delta_{k1} \\
[G_8]_{ik} &= -i\delta_{i1}\delta_{k3} + i\delta_{i3}\delta_{k1} \\
[G_9]_{ik} &= \delta_{i1}\delta_{k4} + \delta_{i4}\delta_{k1} \\
[G_{10}]_{ik} &= -i\delta_{i1}\delta_{k4} + i\delta_{i4}\delta_{k1} \\
[G_{11}]_{ik} &= \delta_{i2}\delta_{k3} + \delta_{i3}\delta_{k2} \\
[G_{12}]_{ik} &= -i\delta_{i2}\delta_{k3} + i\delta_{i3}\delta_{k2} \\
[G_{13}]_{ik} &= \delta_{i2}\delta_{k4} + \delta_{i4}\delta_{k2} \\
[G_{14}]_{ik} &= -i\delta_{i2}\delta_{k4} + i\delta_{i4}\delta_{k2} \\
[G_{15}]_{ik} &= \delta_{i3}\delta_{k4} + \delta_{i4}\delta_{k3} \\
[G_{16}]_{ik} &= -i\delta_{i3}\delta_{k4} + i\delta_{i4}\delta_{k3}
\end{aligned} \tag{16.42}$$

Then completing eq. 16.41 using eq. 16.40 we find

$$\begin{aligned}(D_{\ldots\mu}\,\rho)^\dagger D_{\ldots}^{\;\mu}\,\rho &= \partial\rho_1/\partial X^\mu\partial\rho_1/\partial X_\mu + \partial\rho_2/\partial X^\mu\,\partial\rho_2/\partial X_\mu + g_G^{\,2}\rho_1^{\,2}L_\mu\,L^\mu/4 + g_G^{\,2}\rho_2^{\,2}L_{D\mu}\,L_D^{\,\mu}/4 + \\ &+ (g_G/2)^2\rho_1^{\,2}(U_5^{\,2} + U_6^{\,2}) + (g_G/2)^2\rho_2^{\,2}(U_9^{\,2} + U_{10}^{\,2}) + (g_G/2)^2\rho_1^{\,2}(U_{11}^{\,2} + \\ &+ U_{12}^{\,2}) + + (g_G/2)^2(\rho_1^{\,2} + \rho_2^{\,2})(U_{13}^{\,2} + U_{14}^{\,2}) + \; + (g_G/2)^2\rho_2^{\,2}(U_{15}^{\,2} + U_{16}^{\,2})\end{aligned} \tag{16.43}$$

up to total divergences which generate surface terms which we discard and assuming that all fields satisfy the gauge condition

$$\partial U_i^{\mu}/\partial X^{\mu} = 0 \qquad (16.44)$$

Note that there are no mass terms for $U_7(X)$ and $U_8(X)$ as well as $B_{\mu}(X)$ and $B_{D\mu}(X)$ due to our choice of unitary gauge eq. 16.37. Consequently there are four massless long range fields and 12 gauge fields that acquire masses of three different values: $(g_G/2)\rho_{10}$, $(g_G/2)\rho_{20}$, and $(g_G/2)(\rho_{10}^{\ 2} + \rho_{20}^{\ 2})^{1/2}$ where ρ_{10} and ρ_{20} ar the vacuum expectation values of ρ_1 and ρ_2 respectively. The fields $U_7(X)$ and $U_8(X)$ are not "diagonal" and thus appear in the fermion sector as terms connecting fermions in different generations within the four species of normal fermions and within the four species of Dark fermions.[12] Therefore they do not change the values of any of the four types of particle numbers.

Based on the estimate of eq. 15.6 the ultra weak value of the coupling constant is

$$g_G = (4\pi\alpha_B)^{1/2} \approx 1.218\ (Gm_H^{\ 2})^{1/2} \qquad (16.45)$$

The ultra weak value of the coupling constant implies that the baryonic force with gauge field $B_{\mu}(X)$, which is now part of a quadruplet of fields. It is a massless, long range field that corresponds to that of chapter 15 with the exception that chapter 15 looks ahead to later chapters where we discuss a 16-dimensional space that we call the *Megaverse* in which our universe resides where the baryonic force and force associated with the Dark Baryon force exist beyond our universe and act with other possible "island" universes. (The leptonic and Dark leptonic forces are short range and thus do not extend beyond our universe.)

The two non-diagonal long range forces, being between different generations of a species and having an ultra weak coupling are not of great consequence because of the short lifetime of the higher generations of a species. Therefore, despite their long range, they have only the "shortest" time to exert an inter-generation force before a higher generation particle decays.

Since we expect the other massive fields to have very large masses (and thus very large Higgs field vacuum expectation values) and ultra weak coupling they are not likely to be experimentally found for the foreseeable future.

16.4 Impact of the Generation U(4) Higgs Mechanism on Fermion Generation Masses

The fermion masses of the charged lepton, and the up-type quark, and down-type quark species' generations all show a rapid increase of mass with the generation. For example the u quark mass is a few MeV while the t quark (third generation) has a mass of about 170 GeV/c. The ratio of these masses is about 170,000. While one can account for this great difference by the judicious choices of Higgs' parameter values, when one considers the generational group and its associated numerical quantities: ultra weak coupling, very large U particle masses –

[12] Neutral lepton, charged lepton, up-type quark and down-type quark plus the four corresponding Dark species..

perhaps of the order of hundreds or thousands of GeV/c, and the corresponding very large Higgs particle vacuum expectation values in this U gauge field sector[13] then the differences in fermion masses within a species become more understandable and natural from a Leibniz Principle perspective.

Thus the popular view that the ElectroWeak gauge field symmetry breaking is solely via ElectroWeak Higgs fields is not part of our Extended Standard Model unless the U(4) sector is removed. In our model there are two sets of contributions to fermion symmetry breaking: ElectroWeak Higgs particles symmetry breaking, and Generation group U(4) Higgs particles symmetry breaking. The Generation group causes each species to break into four generations.

In the conventional Standard Model the breakup of species into generations is inserted "by hand." It is not a consequence of the existence of SU(2)⊗U(1) symmetry or symmetry breaking. In our approach the U(4) Generation group causes the appearance of generations. We base the existence of the Generation group[14] on the four conserved particle numbers. Leibniz' Principle and Ockham's Razor then lead to the above construction/derivation.

16.5 Generation Group Higgs Mechanism for Fermion Masses

We now consider the Generation group Higgs Mechanism for the eight species of fermions (four species of "normal" matter and four species of Dark Matter). We shall consider the mass terms for the four normal species, which is the same as that of the four Dark species except for the values in the various species mass matrices. Therefore we define the initial 4-vector for the generations of the normal species by

$$\Psi_s = \begin{bmatrix} \Psi_{11} \\ \Psi_{12} \\ \Psi_{13} \\ \Psi_{14} \\ \cdots \\ \Psi_{41} \\ \Psi_{42} \\ \Psi_{43} \\ \Psi_{44} \end{bmatrix} \qquad (16.46)$$

[13] They are not the Higgs particles of the SU(2)⊗U(1) ElectroWeak sector.

[14] In earlier books we suggested the fermion generations might be the result of a wormhole to another 4-dimensional universe. The new approach is simpler and more consistent with known facts – thus more consistent with the Leibniz Minimax Principle.

where ψ_{ki} is the generation index for the ith generation of the kth species. ψ_{k1} is the wave function for the 1st generation, ψ_{k4} is the 4th generation member of the kth species, and we omit other indices in the interests of clarity. The normal fermion species are ordered charged lepton (k = 1), up-type quark, neutral lepton, and down-type quark (k = 4). Other indices of these wave functions are surpressed in the interests of clarity. A 4th generation fermion of any species is yet to be found experimentally. The lagrangian density mass term for the four normal fermion species is

$$\mathcal{L}_U{}^{UHiggs}{}_{FermionMasses} = \Sigma_{\alpha,\beta}\, \psi_{kL\alpha}\, \eta m_{k\alpha\beta}\, \psi_{kR\beta} + \text{c.c.} \tag{16.47}$$

where $m_{k\alpha\beta}$ is complex constant matrix, and where α, β = 1, … , 4. The total *normal* fermion lagrangian mass terms are

$$\mathcal{L}^{Higgs}{}_{FermionMasses} = \mathcal{L}_U{}^{UHiggs}{}_{FermionMasses} + \mathcal{L}_{EW}{}^{Higgs}{}_{FermionMasses} \tag{16.48}$$

where $\mathcal{L}_{EW}{}^{Higgs}$ is the contribution of ElectroWeak Higgs Mechanism to the fermion masses. Using the vacuum expectation value of η in eq. 16.37 we find

$$\mathcal{L}_U{}^{UHiggs}{}_{FermionMasses} = \Sigma_{\alpha,\beta}\, \{\psi_{2L\alpha}\, \rho_1 \overline{m}_{2\alpha\beta}\psi_{2R\beta} + \psi_{4L\alpha}\, \bar{\rho}_2 m_{4\alpha\beta}\psi_{4R\beta}\} + \text{c.c.} \tag{16.49}$$

giving mass terms for the up-type and down-type quark species but not for lepton species. There is an implicit color summation over the color quarks in each generation and quark species. Qualitatively eq. 16.49 could be viewed as corresponding to the experimentally determined largeness of quark masses relative to lepton masses in each generation of normal matter.

The mass matrices $m_2 = [m_{2\alpha\beta}]$ and $m_4 = [m_{4\alpha\beta}]$ are both complex, constant mass matrices. They can be brought to diagonal form with non-negative values by U(4) matrices A_k and B_k:

$$\begin{aligned} A_2 m_2 B_2{}^{-1} &= D_2 \\ A_4 m_4 B_4{}^{-1} &= D_4 \end{aligned} \tag{16.50}$$

or

$$\begin{aligned} m_2 &= A_2{}^{-1} D_2\, B_2 \\ m_4 &= A_4{}^{-1} D_4\, B_4 \end{aligned} \tag{16.51}$$

We now note, that although, both D_2 and D_4 have non-negative real values, down-type quarks are all tachyonic and up-type quarks are all non-tachyonic due to their lagrangian kinetic terms as seen in chapter 5.

We further note that $m_2^\dagger m_2$ and $m_4^\dagger m_4$ are hermitean, and A_k and B_k are members of U(4) as is D_k for k = 2,4, with the result that m_2 and m_4 are also both members of the U(4) group. Thus

$$m_2^{-1} = m_2^\dagger \qquad (16.52)$$
$$m_4^{-1} = m_4^\dagger$$

We can express the mass matrices in terms of U(4) generators

$$m_2 = \Sigma G_i m_{2i} \qquad (16.53)$$
$$m_4 = \Sigma G_i m_{4i}$$

$$m_2^{-1} = m_2^\dagger = \Sigma G_i m_{2i}^* \qquad (16.54)$$
$$m_4^{-1} = m_4^\dagger = \Sigma G_i m_{4i}^*$$

since the matrices G_i are all hermitean, where $\{m_{2i}\}$ and $\{m_{4i}\}$ are each a set of sixteen complex constants.

While we do not as yet know the 4th generation fermions or their masses, the third generation quarks have masses that are far greater than the 1st and 2nd generation quarks or their sum suggesting that the trace of m_2 and m_4.is dominated by the 4th generation mass of the two quark species with a similar situation holding, perhaps, for the two Dark quark species. Therefore if we take the trace of m_2 and m_4 then it seems probable based on the trend of the generations that the 4th generation mass dominates the trace:

$$D_{24} \approx \text{tr}\, D_2 \qquad (16.55)$$
$$D_{44} \approx \text{tr}\, D_4$$

We can use these A_k and B_k U(4) transformations to define the eight "physical" (up to further ElectroWeak Higgs Mehanism effects) up-type and down-type quark generations fields:

$$\psi_{2L\alpha}\, \rho_1 m_{2\alpha\beta}\psi_{2R\beta} + \psi_{4L\alpha}\, \rho_2 m_{4\alpha\beta}\psi_{4R\beta} = (\psi_{2L}\, A_2^{-1})_\alpha\, \rho_1 D_{2\alpha\beta}(B_2\psi_{2R})_\beta + (\psi_{4L}\, A_4^{-1})_\alpha\, \rho_2 D_{4\alpha\beta}(B_4\psi_{4R})_\beta$$
$$= \psi_{2Lphys\alpha}\, \rho_1 D_{2\alpha\beta}\psi_{2Rphys\beta} + \psi_{4Lphs\alpha}\, \rho_2 D_{4\alpha\beta}\psi_{4Rphys\beta} \qquad (16.56)$$

Species: up-type quarks down-type quarks

Note that eq. 16.56 show that lepton masses do not have U(4) Higgs Mechanism contributions.

The preceding discussion with changes in the values of constants and constant matrices holds for Dark Matter also where the Dark quarks acquire mass terms, but the Dark leptons do not, *by the Dark equivalent of eq. 16.56.* The Dark Matter species mass terms, with the subscript D signifying Dark Matter, are

$$= \psi_{D2Lphys\alpha}\, \rho_{D1} D_{D2\alpha\beta} \psi_{D2Rphys\beta} + \psi_{D4Lphs\alpha}\, \rho_{D2} D_{D4\alpha\beta} \psi_{D4Rphys\beta} \qquad (16.57)$$

Dark Species: up-type quarks down-type quarks

The Dark Higgs bosons' vacuum expectation values are not required to be the same as the "normal" Higgs bosons' vacuum expectation values, and perhaps may contribute to large Dark quark masses.

3. Mass, Renormalization, Higgs Fields, and Symmetries in the Extended Standard Model

3.1 Why Does the Standard Model Require Massless Gauge and Fermion fields?

Since the proof of the renormalizability of ElectroWeak theory by t'Hooft, Veltman and collaborators the basis of the theory in massless fermions and vector bosons, and initially unbroken SU(2)⊗U(1) that subsequently is broken by the Higgs Mechanism has largely gone unchallenged. Renormalization removed infinities that might otherwise appear in ElectroWeak perturbation theory calculations.

In volume I and earlier books we showed another approach existed, Two-Tier Quantum Field Theory, based on a form of quantum coordinates that eliminated all infinities that might otherwise appear in ElectroWeak, and other quantum field theories, perturbation theory calculations – Feynman diagram by Feynman diagram. This formulation eliminated the need for the t'Hooft approach, and other approaches, for renormalization to eliminate infinities.

3.2 Finiteness of the Extended Standard Model

Using a Two-Tier QFT formulation of The Standard Model and our Extended Standard Model we can eliminate all divergences (infinities) diagram by diagram. There is no need for an unbroken symmetry unlike t'Hooft's approach. Fermions can have masses initially that may change due to interactions – but in a finite amount. Gauge bosons can also have masses although the gauge symmetries are violated *ab initio.*

The Two-Tier formalism also enables higher dimension Two-Tier quantum field theories to be created that are finite as we did in volume I in the case of our 16 dimension Megaverse.

In volume I we kept the spontaneous breakdown Higgs Mechanism formalism for fermions and vector bosons. Our motivation was two-fold. First, we wanted to show that the Extended Standard Model could have the same Standard Model sector as had been previously portrayed in Physics Literature. Secondly, it appeared likely that scalar Higgs particles would be found experimentally. These particles appear to have been found[15] at the CERN LHC, and they are associated with W and Z gauge bosons in the manner that conforms to the predictions of The Standard Model, and our Extended Standard Model as portrayed in volume I and earlier books.

[15] S. Chatrtchyan et al, Phys. Rev. D **89**, 012003 (2014) and references therein.

Thus the Extended Standard Model is in agreement with the Higgs features found at LHC.

3.3 Iotas and Symmetry Breakdown in the Extended Standard Model

Remarkably, our Extended Standard Model which is an attempt at a deeper theory, based on the only aspect of reality that must be true in any Physics theory, namely Logic, does not begin with gauge boson symmetries – since all fermions have a component, the *iota*, which embodies 4-valued Asynchronous Logic. The iota has an extremely small mass, the Landauer mass, which represents the minimal energy with a quantum of information. In our derivation, the 4-valued iotas, *with mass*, require 4-dimensional space-time dimensionality, and inexorably lead to four fermion species: charged leptons, neutral leptons, up-type quarks and down-type quarks.

Yet the iota mass embedded in fermions causes the breakdown of ElectroWeak SU(2)⊗U(1) symmetry making a t'Hooft proof of renormalizability impossible.

Thus Two-Tier quantum field theory becomes a necessity since gauge symmetry is not present and not required for renormalizability. The finiteness of perturbation theory is guaranteed without recourse to symmetry-based approaches.

3.4 Why are Higgs Fields present in the Extended Standard Model?

Since our Extended Standard Model does not need exact symmetries with massless particles and, in fact, violates SU(2)⊗U(1)symmetry due to the iota mass, why has nature bestowed Higgs particles on us? One could imagine a finite Standard Model using Two-Tier quantization that would nicely avoid infinities in perturbation theory.

So it appears that we cannot justify the necessity of Higgs particles by the requirement of finite renormaizability.

Only one justification is apparent that may prove to be the purpose of Higgs particles. It may be that the appearance of Higgs particles to generate particle masses may be a hint of a larger role, in which constants appearing in the Extended Standard Model such as particle masses and coupling constants may all result from the vacuum expectation values of Higgs particles in a deeper, as yet undiscovered, theory. Some time ago Dicke suggested that the gravitational coupling constant G may be the consequence of "the average value" (vacuum expectation value) of a scalar field.[16] Perhaps the constants of the Extended Standard Model may also reflect vacuum expectation values of Higgs scalars. Then the success in deriving the form of The Extended Standard Model from first principles may be further extended to a derivation of the constants appearing in The Extended Standard Model, thus providing a deeper rationale for Higgs scalars. These constants must be determined from first principles.

[16] See Weinberg () p. 155ff for a lucid discussion.

4. A Grand Unified U(4)⊗U(4) (GUT) Theory

Studies of the change of the effective coupling constants of the three experimentally known interactions have suggested that the values of the coupling constants may become equal at very high energies and support unification of the interactions into a Grand Unified Theory (GUT).[17] Our Extended Standard Model introduces a new Dark ElectroWeak SU(2)⊗U(1) interaction enlarging the symmetry to SU(3)⊗SU(2)⊗U(1)⊗SU(2)⊗U(1)⊗U(4). It appears reasonable to believe the two new interaction coupling constants will be similar in value to the known Electroweak coupling constants and may join the three known coupling constants to unify at some large energy of perhaps $M_{GUT} \approx 10^{17}$ GeV.

Since U(4) contains the non-commuting subgroups: SU(3), SU(2), U(1), SU(2), and U(1) it is also reasonable, and in accord with Ockham's Razor, to assume the Extended Standard Model symmetries join to eventually form a U(4) symmetry at the anticipated unification energy. This choice is further supported by the appearance of U(4) as the Reality group of complex General Relativity, and also the role of U(4) in the creation of a Theory of Everything, described in later chapters.

Taking U(4) as the unification group, and combining it with the U(4) Generation group, we find the GUT symmetry group to be

$$U(4)\otimes U(4) \tag{3.1}$$

As noted in the preceding chapter the vast disparity between the coupling constants of the strong and EletroWeak interactions vs. the Generation group coupling constant suggest that a further unification, at perhaps a much higher energy, is not likely.

A possible reason for the failure to unify the U(4) sectors is the differing origins of each sector. The unified U(4) originates in the Reality group SU(3)⊗SU(2)⊗U(1)⊗SU(2)⊗U(1) which is based on complex space-time. The Generation group U(4) originates in the four conserved particle number symmetries as seen in the preceding chapter. Consequently the vast difference in coupling constants is understandable. Since the gravitational constant G and the approximate value of the baryonic coupling constant (and thereby the U(4) coupling constant) of $0.118Gm_H^2$ are comparable it appears more reasonable to consider a unification of the Gravitation coupling constant with the Generation coupling constant.

GUT theories of the form U(N)⊗U(M) for N, M ranging up to three have been previously considered.[18] However they have not been related to complex space-time and particle number symmetries as we have derived in these volumes and earlier.

Thus we have a Grand Unified Theory based on U(4)⊗U(4) if coupling constants unify at some large energy.

---

[17] H. Georgi and S. Glashow, Phys. Rev. Lett., **32**, 438 (1974), U. Amaldi, W. de Boer and H. Furstenau, Phys. Lett. **B260**, 447 (1991).

[18] B. Gato-Rivera and A. N. Schellekens, arXiv: 1401.1782 (2014) and references therein.

5. New Formulation of Complex General Relativity

We know that Special Relativity covariant equations go directly over to General Relativity covariant equations.[19] Similarly General Relativity smoothly passes directly to Complex General Gelativity by analyitc continuation.

In Blaha (2004) and volume I we considered the General Relativity of complex-valued space-time. We pointed out that Complex General Relativity can be obtained from the General Relativity of real-valued space-time coordinates by piece-wise analytic continuation using complex variable theory. We then constructed Complex General Relativity and considered some new features that appeared which are described in the appendices 5-A and 5-B of this chapter and in Blaha (2004).

In this chapter we extend our discussion of Complex General Relativity *to a new form* – to include the Reality group that we used in the flat space-time derivation of the Extended Standard Model.

The analytic continuation of coordinates to complex values yields a Complex General Relavity. Another way of creating a Complex General Relativity is to apply Reality group transformations to real-valued coordinates to transform them to complex values. In this approach a local Reality group transformation is applied to real-valued coordinates. The Reality group transformation is a linear transformation with complex local coefficients. The result is a complex, local coordinate system. This approach differs from analytic continuation. Analytic continuation complexifies each of the four coordinates of a real-valued General Relativistic solution without mixing the coordinates. A Reality group transformation "mixes" the coordinates using complex coefficients.

The result is a different formulation of Complex General Relativity that we will see somewhat resembles the formulation of the Extended Standard Model using the Reality group. This formulation introduces a new set of scalar fields[20,21] Φ_k for k = 1, 2, … , 16 whose dynamic equation has kinetic terms which we will take to be the kinetic derivative terms of Higgs boson dynamic equations. These kinetic terms, plus potential terms that implement the Higgs Mechanism, will be taken to be part of the Higgs sector of the Extended Standard Model. Thus Complex Gravity is united with the Extended Standard Model through the scalar particles required by the Reality group version of Complex General Relativity and the Higgs Mechsnism. *Together they become a Theory of Everything.*

---

[19] Weinberg (1972).

[20] These scalar fields have dynamic equations that greatly differ from the scalar fields in Dicke-Brans-Jordan theories of Gravitation.

[21] The introduction of gauge field Reality transformations, rather than scalar fields, would lead to unacceptable first order gauge field dynamic equations.

Consequently, we chose to provisionally introduce the features of a new Complex General Relativity using the Reality group to formulate its features, and to modify its dynamics, so as to relate Complex General Relativity directly to the Extended Standard Model and thus lead to a true Theory of Everything. (Chapter 6)

5.1 Complex General Relativity

Complex General Relativity like real General Relativity has its origin in 4-dimensional space-time. In chapters 2 and 4 of volume I we showed that 4-dimensional space-time had its origin in Asynchronous Logic. We then showed that flat space-time was complex. Generalizing to curved space-time it is necessary for consistency to assume a *complex* curved space-time that is made real using the Reality group, and – applying Ockham's Razor and Leibniz's Minimax Principle – the simplest choice yielding compatibility between complex flat space-time and its curved space-time extension.

We can then derive the Einstein dynamic equations of General Relativity for both real-valued and complex-valued coordinates. Appendices 5-A and 5-B outline the derivation of Complex General Relativity by analytic continuation, and some of its features. The similarity to real-valued General Relativity is apparent (despite some new features), as it must, due to piece-wise analytic continuation. The interested reader may wish to read appendix 5-A before proceeding to the next section.

5.2 Reality Group Transformations in Curved Complex Space-Time

The U(4) Reality group enables any 4-dimensional complex vector to be transformed to a real-valued vector or *vice versa*. If we define local Reality group transformations then we can transform the set of coordinates of a reference frame to real values. In this section we will specify the matrix form of the generators of U(4) and relate them to the U(4) subgroups. Then we will define tetrad (vierbein) forms of the generators in an inertial reference frame for later use in this chapter in defining the dynamic equations of the new Complex General Relativity.

It will become apparent that the use of the Reality group to map Complex General Relativity to real General Relativity will reveal a correspondence with the Extended Standard Model that we will use in chapter 6 to establish a Theory of Everything.

The Extended Standard Model sector can then be viewed as being defined in an inertial reference frame at each point in a curved space-time. If we imagine the curved space-time as smoothly beconing flat then the curved space-time definition of the Extended Standard Model then extends throughout flat space-time.

5.2.1 4-Dimensional Representation of the Subgroup Generators of the U(4) Reality Group

Four dimensional representations of some of the sixteen generators of U(4) were defined in Blaha (2011c). These generators can be put in the form of the non-commuting subgroups' 16 generators for SU(3), SU(2), U(1), SU(2) and U(1). Since we wish to use these generators to implement transfomations in 4-dimensional space-time we must define 4×4 matrix *reducible* representations of each subgroup's set of generators. Thus the sixteen 4×4 generator τ_k (for k = 1, … , 16) representations are:

SU(3) – a $\underline{3}\oplus\underline{1}$ SU(3) representation of 8 generators
SU(2) – a $\underline{3}\oplus\underline{1}$ SU(2) representation of 3 generators
U(1) – a $\underline{1}\oplus\underline{1}\oplus\underline{1}\oplus\underline{1}$ U(1) representation of 1 generator
SU(2) – another $\underline{3}\oplus\underline{1}$ SU(2) representation of 3 generators
U(1) – another $\underline{1}\oplus\underline{1}\oplus\underline{1}\oplus\underline{1}$ U(1) representation of 1 generator

5.2.2 Scalar Fields of the Reality group

Since we will develop a unified theory of everything later we will define scalar fields using the notation of the Extended Standard Model in volume I:

$$\begin{aligned} &SU(3) - \Phi_i \text{ for } i = 1, 2, \ldots, 8 \\ &SU(2) - \Phi_i \text{ for } i = 1, 2, 3 \\ &U(1) - \Phi_0 \\ &SU(2) - \Phi'_i \text{ for } i = 1, 2, 3 \\ &U(1) - \Phi'_0 \end{aligned} \tag{5.1}$$

It is convenient to relabel the above fields as $\Phi_k(x)$ for k = 1, 2, …, 16. It will also be convenient to label the 4×4 generator matrices as listed above as τ_k in the same respective order.

The combined treatment of the various Reality group symmetries in a 4×4 matrix representation may seem at first strange. However their combination within one representation (based on the four dimensionality of space-time) for use in defining a part of the complex general relativistic transformations of Complex General Relativity is acceptable and, more importantly, provides the needed complexity of the theory as will be seen. We note that the transformation defined below in eq. 5.2 below is a matrix but the symmetries of the various fields $\Phi_k(x')$ is subsumed in the inner product with U(4) generator matrices τ_k and so U(x), defined below, becomes simply a broken U(4) 4×4 matrix dependent on x.

The U(4) generators may be used to define a broken U(4) general coordinate transformation in four dimensions which corresponds to the Reality group:[22]

$$U(x) = \exp[i \textstyle\sum_k g_k\Phi_k(x')\tau_k] \equiv e^{i\theta(x)} \tag{5.2}$$

[22] If there is a GUT form of the Extended Standard Model as we considered in the previous chapter, then all coupling constants g_i may become equal at the GUT energy and the transformation will be an unbroken U(4) transformation.

where the sum is for k = 1, ..., 16. The set of coupling constants, $\{g_k\}$, consist of repetitions of the five different interactions' coupling constants for the five subgroups listed above.

5.2.3 Tetrad (Vierbein) Form of the 4×4 Reality Group Transformations

In this subsection we will map 4×4 Reality group transformations (eq. 5.2) to the form of General Relativistic transformations using *tetrads* (*vierbeins*). Then we will construct Complex General Relativistic transformations, the complex affine connection, and then dynamical gravitational equations.

The *vierbein* formalism begins with the Equivalence Principle that allows us to define an inertial coordinate system in the neighborhood of any point Z in space-time. We will use the notation $\varsigma^\alpha(Z)$ to denote the inertial coordinates at Z. We define a tetrad or vierbein as

$$v^\alpha{}_\mu(x) = (\partial\varsigma^\alpha(x)/\partial x^\mu)_{x=Z} \qquad (5.3)$$

and, in a neighborhood of Z, we can invert the relation between ς and x to define an inverse

$$w^\mu{}_\alpha(x) = (\partial x^\mu(\varsigma)/\partial\varsigma^\alpha)_{x=X} \qquad (5.4)$$

such that

$$w^\mu{}_\alpha(x)v^\alpha{}_\nu(x) = \delta^\mu{}_\nu \qquad (5.5)$$
$$w^\mu{}_\beta(x)v^\alpha{}_\mu(x) = \delta^\alpha{}_\beta \qquad (5.6)$$

In real General Relativity all *tetrads* are real-valued. In Complex General Relativity a *tetrad* $v^\alpha{}_\mu(x)$ is complex-valued.

The metric at a curved space-time point X is defined in terms of *tetrads* as

$$g_{\rho\sigma}(x) = \eta_{\alpha\beta}\, v^\alpha{}_\rho(x)v^\beta{}_\sigma(x) \qquad (5.7)$$
$$g^{\rho\sigma}(x) = \eta^{\alpha\beta}\, w^\rho{}_\alpha(x)w^\sigma{}_\beta(x) \qquad (5.8)$$

The inverse of a *tetrad* transformation can also be expressed as

$$w_\beta{}^\nu(x) = v_\beta{}^\nu(x) = \eta_{\beta\alpha}g^{\nu\mu}(x)v^\alpha{}_\mu(x) \qquad (5.9)$$

Then a *tetrad* and its inverse satisfy the relations

$$v^\alpha{}_\mu(x)v_\beta{}^\mu(x) = \delta^\alpha{}_\beta \qquad (5.10)$$

and

$$v^\alpha{}_\mu(x)v_\alpha{}^\nu(x) = \delta^\nu{}_\mu \qquad (5.11)$$

5.2.3.1 Transformations of Tetrads

There are two general types of space-time transformations that can be performed on a tetrad.

1. A complex-valued (possibly real-valued) General Relativistic coordinate transformation:

$$v'^{\alpha}{}_{\mu}(x) = \partial x^{\nu}/\partial x'^{\mu}\, v^{\alpha}{}_{\nu}(x) \tag{5.12}$$

2. A complex-valued, local *Lorentzian transformation*

$$v'^{\beta}{}_{\mu}(x) = \Lambda(x)^{\beta}{}_{\alpha}\, v^{\alpha}{}_{\mu}(x) \tag{5.13}$$

where $\Lambda(x)^{\beta}{}_{\alpha}$ is an element of a subset of the local Complex Lorentz Group to be specified later.

5.2.3.2 Local Lorentzian Formalism for Tetrads

The local Lorentzian transformations $\Lambda(x)^{\beta}{}_{\alpha}$ consist of local Lorentz transformations that are real-valued, and complex-valued Lorentz transformations. Both types of transformations satisfy the orthogonality condition:

$$\eta_{\alpha\beta}\Lambda^{\alpha}{}_{\rho}(x)\Lambda^{\beta}{}_{\sigma}(x) = \eta_{\rho\sigma} \tag{5.14}$$

Thus the *tetrad* partakes of both local (position dependent) General Relativistic transformations and local Lorentzian transformations.

5.2.3.3 Tetrad (Vierbein) Form of the 4×4 Reality Group Transformations

In eq. 5.2 we defined Reality group matrix transformations. Using the matrix form of this definition, and tetrads, we see that we can express a Reality transformation in an inertial coordinate system in the neighborhood of a point Z in space-time. Thus

$$\begin{aligned} U^{\mu}{}_{\nu}(x) &\equiv w^{\mu}{}_{a}(x)U^{a}{}_{b}(x)v^{b}{}_{\nu}(x) = w^{\mu}{}_{a}(x)[e^{i\theta(x)}]^{a}{}_{b}v^{b}{}_{\nu}(x) \\ &= w^{\mu}{}_{a}(x)[\textstyle\sum_{n} (i\theta(x))^{n}/n!]^{a}{}_{b}v^{b}{}_{\nu}(x) \\ &= w^{\mu}{}_{a}(x)v^{a}{}_{\nu}(x) + w^{\mu}{}_{a}(x)i\theta(x)^{a}{}_{b}v^{b}{}_{\nu}(x) + \ldots \\ &= \delta^{\mu}{}_{\nu} + [\theta(x)]^{\mu}{}_{\nu} + [\theta(x)]^{\mu}{}_{\alpha}[\theta(x)]^{\alpha}{}_{\nu}/2 + \ldots \end{aligned} \tag{5.15}$$

where we have transformed the terms within the expansion of the exponentiated matrix into local Lorentz frame tensors, and where

$$[\theta(x)]^{\mu}{}_{\alpha} = w^{\mu}{}_{a}(x)\, [\theta(x)]^{a}{}_{b}\, v^{b}{}_{\alpha}(x) \tag{5.15a}$$

with a and b being the 4×4 matrix column and row indices of $[\tau_k]^a{}_b$ (See eq. 5.2.). The local Lorentzian tensorial form of U(x) can be used to develop Complex General Coordinate transformations as we do in the next section.

For later use we note that the inverse of eq. 5.15 is

$$U^{-1\alpha}{}_\mu(x) = w^\alpha{}_a(x)U^{-1a}{}_b(x)v^b{}_\mu(x) \quad (5.16)$$

with

$$U^{-1\alpha}{}_\mu(x)U^\mu{}_\nu(x) = \delta^\alpha{}_\nu \quad (5.17)$$

where

$$U^{-1a}{}_b(x) = [U^\dagger(x)]^a{}_b \quad (5.18)$$

with † signifying hermitean conjugate.

5.3 Structure of Complex General Coordinate Transformations

Complex General Coordinate transformations can be uniquely factored into products of two terms. They have the form

$$\partial x''^\nu(x)/\partial x^\mu = U(x'')^\nu{}_\beta\, \partial x'^\beta(x)/\partial x^\mu \quad (5.19)$$

where

$$x''^\nu(x) = U(x'')^\nu{}_\beta x'^\beta$$
$$x'^\mu(x) = U^{-1\mu}{}_b(x'')\, x''^b$$

where $U(x')^\nu{}_\beta$ is complex and where $\partial x'^\beta(x)/\partial x^\mu$ is a purely real General Coordinate transformation.

We define

$$U(x'')^\mu{}_\nu = w^\mu{}_a(x'')\big[\exp\big(i \textstyle\sum_k g_k\Phi_k(x'')\tau_k\big)\big]^a{}_b\, v^b{}_\nu(x'') \quad (5.20)$$
$$U^{-1}(x'')^\mu{}_\nu = w^\mu{}_a(x'')\big[\exp\big(-i\textstyle\sum_k g_k\Phi_k(x'')\tau_k\big)\big]^a{}_b\, v^b{}_\nu(x'') \quad (5.21)$$

where the constants g_k are real, and Φ_k and τ_k are hermitean. The uniqueness of the factorization follows from the Reality group (and U(4)) property that any complex 4-vector can be uniquely mapped to any specified real 4-vector.

Given the factorization (eq. 5.19) it becomes possible to separate the affine connection correspondingly. Then the dynamical gravitational equations of Complex General Relativity can be made to exhibit their Φ_k field dependence in a manner analogous to Higgs fields in the Extended Standard Model. And a Theory of Everything becomes evident. (See chapter 6.)

5.4 Complex Affine Connection

The structure of a complex general coordinate transformation (eq. 5.19) enables us to calculate its affine connection for later use in determining the covariant derivative, and the dynamic equations. First the transformation to the real-valued x' coordinates from inertial coordinates is

$$\Gamma^{\sigma}{}_{\lambda\mu}\,(x') = \partial x'^{\sigma}/\partial\varsigma^{\rho}\ \partial^{2}\varsigma^{\rho}/\partial x'^{\lambda}\partial x'^{\mu} \qquad (5.22)$$

Next the Reality group transformation has the affine connection

$$\Gamma^{\sigma}{}_{\lambda\mu}\,(x'') = \partial x''^{\sigma}/\partial\varsigma^{\rho}\ \partial^{2}\varsigma^{\rho}/\partial x''^{\lambda}\partial x''^{\mu} \qquad (5.23)$$

which becomes

$$\Gamma^{\sigma}{}_{\lambda\mu}\,(x'') = \partial x''^{\sigma}/\partial x'^{\beta}\ \partial x'^{\beta}(\varsigma)/\partial\varsigma^{\rho}\ \partial/\partial x''^{\mu}[\partial\varsigma^{\rho}/\partial x'^{\alpha}\ \partial x'^{\alpha}/\partial x''^{\lambda}] \qquad (5.24)$$

Using eq. 5.22 we find eq. 5.24 has the form

$$\Gamma^{\sigma}{}_{\lambda\mu}(x'') = \partial x''^{\sigma}/\partial x'^{\beta}\ \partial x'^{\alpha}/\partial x''^{\lambda}\ \partial x'^{\gamma}/\partial x''^{\mu}\ \Gamma^{\beta}{}_{\alpha\gamma}(x') + \partial x''^{\sigma}/\partial x'^{\beta}\ \partial^{2}x'^{\beta}/\partial x''^{\lambda}\partial x''^{\mu} \qquad (5.25)$$

Next substituting the Reality group transformation

$$x''^{\nu}(x) = U(x'')^{\nu}{}_{\beta}x'^{\beta} \qquad (5.26)$$
$$x'^{\mu}(x) = U^{-1}(x'')^{\mu}{}_{\beta}\,x''^{\beta}$$

together with

$$\partial x''^{\sigma}/\partial x'^{\beta} = \partial[U(x'')^{\sigma}{}_{\alpha}x'^{\alpha}]/\partial x'^{\beta} = U(x'')^{\sigma}{}_{\beta} + x'^{\alpha}\,\partial U(x'')^{\sigma}{}_{\alpha}/\partial x'^{\beta} \qquad (5.27)$$
$$\partial x'^{\sigma}/\partial x''^{\beta} = \partial[U^{-1}(x'')^{\sigma}{}_{\alpha}x''^{\alpha}]/\partial x''^{\beta} = U^{-1}(x'')^{\sigma}{}_{\beta} + x''^{\alpha}\,\partial U^{-1}(x'')^{\sigma}{}_{\alpha}/\partial x''^{\beta} \qquad (5.28)$$

we find the second term in eq. 5.25 is the Reality fields affine connection

$$\Gamma_{V}{}^{\sigma}{}_{\lambda\mu}(x'') = \partial[U(x'')^{\sigma}{}_{\alpha}x'^{\alpha}]/\partial x'^{\beta}\ \partial\{\partial[U^{-1}(x'')^{\beta}{}_{\alpha}x''^{\alpha}]/\partial x''^{\lambda}\}/\partial x''^{\mu} \qquad (5.29)$$

5.5 Complex Curvature Tensor and Complex Einstein Equation of the New Complex General Relativity

The complex space-time Riemann-Christoffel curvature tensor is

$$R^{\rho}{}_{\mu\nu\sigma}(x'') \equiv \partial\Gamma^{\rho}{}_{\mu\nu}(x'')/\partial x''^{\sigma} - \partial\Gamma^{\rho}{}_{\mu\sigma}(x'')/\partial x''^{\nu} + \Gamma^{\alpha}{}_{\mu\nu}(x'')\Gamma^{\rho}{}_{\sigma\alpha}(x'') - \Gamma^{\alpha}{}_{\mu\sigma}(x'')\Gamma^{\rho}{}_{\nu\alpha}(x'') \qquad (5.30)$$

as can be shown by following the standard steps of its derivation. (We note that the algebra of the tensor manipulations is the same as in the usual derivation of General Relativistic quantities and equations. The features of the curvature tensor, related quantities, and other expresions are listed in appendix 5-A.)

The Complex General Relativistic Einstein Equation is

$$R_{\mu\nu}(x'') - \tfrac{1}{2}\, g_{\mu\nu}\, R(x'') = -8\pi G\, T_{\mu\nu} \tag{5.31}$$

where G is Newton's gravitational constant (5.674×10^{-11} $m^3kg^{-1}s^{-2}$), $T_{\mu\nu}$ is the energy-momentum tensor, the Ricci tensor is

$$R_{\mu\nu}(x'') = R^{a}{}_{\mu a\nu}(x'') \tag{5.32}$$

and the curvature scalar is

$$R(x'') = g^{\mu\nu}R_{\mu\nu}(x'') \tag{5.33}$$

with $g^{\mu\nu} = g^{\mu\nu}(x'')$.

5.6 Approximate Form of the New Complex General Relativity Dynamic Equations

The introduction of scalar fields to embody the transformation of real-valued coordinates and expressions to complex values creates a new form of Complex General Relativity if the scalar fields' have dynamic equations, which together with the complex Einstein equation, are jointly solved. A solution of the complete set of equations is not possible presently.

Since the form of $\Gamma^{\sigma}{}_{\lambda\mu}(x'')$ as given in eqs. 5.25 – 5.29 is sufficiently complicated to make a solution of the new Einstein equation eq. 5.31 impossible currently, we will explore an approximete solution. It is possible to derive a leading order approximation if the scalar fields are sufficiently weak (with the result that the complex gravitational field is almost real-valued), and if the gravitational field is also sufficiently weak. In this case, we can approximate $U(x'')^{\nu}{}_{\beta}$ and $U^{-1}(x'')^{\nu}{}_{\beta}$ of eqs. 5.20-5.21 with

$$U(x'')^{\nu}{}_{\beta} \approx \delta^{\nu}{}_{\beta} + i\sum_k g_k\Phi_k(x'')[\tau_k]^{\nu}{}_{\beta} \tag{5.34}$$

$$U^{-1}(x'')^{\nu}{}_{\beta} \approx \delta^{\nu}{}_{\beta} - i\sum_k g_k\Phi_k(x'')[\tau_k]^{\nu}{}_{\beta} \tag{5.35}$$

using

$$w^{\mu}{}_{a}(x'') \approx \delta^{\mu}{}_{a}$$
$$v^{b}{}_{\nu}(x'') \approx \delta^{b}{}_{\nu}$$

Then eq. 5.25 becomes

$$\Gamma^{\sigma}{}_{\lambda\mu}(x'') \approx \Gamma_{GR}{}^{\sigma}{}_{\lambda\mu}(x') + \Gamma_V{}^{\sigma}{}_{\lambda\mu}(x'') \tag{5.36}$$

to leading order in $V_{\mu k}(x'')$ where $\Gamma_{GR}{}^{\sigma}{}_{\lambda\mu}(x')$ is the usual general relativistic affine connection and where

$$\Gamma_V{}^{\sigma}{}_{\lambda\mu}(x'') \equiv U(x'')^{\sigma}{}_{\beta}\, \partial[U^{-1}(x'')^{\beta}{}_{\lambda}]/\partial x''^{\mu} \tag{5.37}$$
$$\approx -\tfrac{1}{2}\, i\{\sum_k g_k\partial\Phi_k(x'')/\partial x''^{\mu}\, [\tau_k]^{\sigma}{}_{\lambda} + \sum_k g_k\partial\Phi_k(x'')\,/\partial x''^{\lambda}[\tau_k]^{\sigma}{}_{\mu}\}$$

in view of the forms of the Ricci tensor and the scalar curvature – the other terms lead to zero when used in the Einstein dynamic equation. In a more compact notation,

$$\Gamma_V{}^{\sigma}{}_{\lambda\mu}(x'') \approx iA^{\sigma}{}_{\lambda\mu}$$

with $\lambda\mu$ symmetry where

$$A^{\sigma}{}_{\lambda\mu} = -½\{\sum_k g_k\partial\Phi_k(x'')/\partial x''^{\mu}\, [\tau_k]^{\sigma}{}_{\lambda} + \sum_k g_k\partial\Phi_k(x'')\, /\partial x''^{\lambda}[\tau_k]^{\sigma}{}_{\mu}\} \qquad (5.37a)$$

$$A^{\sigma}{}_{\lambda\mu} = -½\{\partial/\partial x''^{\mu}\, \varphi^{\sigma}{}_{\lambda} + \partial\, /\partial x''^{\lambda}\, \varphi^{\sigma}{}_{\mu}\} \qquad (5.37b)$$

with

$$\varphi^{\sigma}{}_{\mu} = \sum_k g_k\Phi_k(x'')[\tau_k]^{\sigma}{}_{\mu} \qquad (5.37c)$$

5. 7 Approximate Particle Motion in a Weak Gravity Field and in Weak Reality Fields Φ_k

The motion of a particle in a freely falling coordinate system ς^{ρ} satisfies

$$d^2\varsigma^{\rho}/d\tau^2 = 0$$

which can be expressed in an arbirary coordinate system x^{ρ} in the form

$$d^2x^{\rho}/d\tau^2 + \Gamma^{\rho}{}_{\mu\nu}\, dx^{\mu}/d\tau\, dx^{\nu}/d\tau = 0$$

In a slowly moving partile in a weak stationary gravitational and in weak scalar fields we can approximate the particle dynamic equation with

$$d^2x^{\rho}/d\tau^2 + \Gamma^{\rho}{}_{00}\, (dt/d\tau)^2 = 0$$

neglecting spatial derivative terms. Substituting eqs. 5.36 and 5.37 we obtain

$$d^2x^{\rho}/d\tau^2 + [\Gamma^{\rho}{}_{00}(x) + i\Gamma_V{}^{\rho}{}_{00}(x)]\, (dt/d\tau)^2 = 0$$

where we approximate x" = x' = x with

$$\Gamma_V{}^{\rho}{}_{00}(x) = A^{\rho}{}_{00}$$

if all the $V_{0k}(x)$ fields are slowly varying with time. Thus

$$d^2x^{\rho}/d\tau^2 + [\Gamma^{\rho}{}_{00}(x) + i\, A^{\rho}{}_{00}](dt/d\tau)^2 = 0$$

The equation for t is

$$d^2t/d\tau^2 + i\, A^{\rho}{}_{00}\, (dt/d\tau)^2 = 0$$

under the assumption that the gravitational field is weak. Its solution is

$$t - t_0 = -\ln(\tau - \tau_0)/\{i\, A^\rho{}_{00}\}$$

assuming the time dependence of $V_{0k}(x)$ is negligible. As a result

$$dt/d\tau = -\, 1/\{i(\tau - \tau_0)A^\rho{}_{00}\}$$

Assuming a weak field approximation so that

$$g_{\mu\nu} \approx \eta_{\mu\nu} + h_{\mu\nu}$$

we see the gravitational affine connection can be approximated with

$$\Gamma^\rho{}_{00}(x) = -\tfrac{1}{2}\, \eta^{\rho\nu}\, \partial h_{00}/\partial x^\nu$$

resulting in the spatial equation

$$d^2\mathbf{x}/d\tau^2 = \tfrac{1}{2}\, (dt/d\tau)^2\, \nabla\, h_{00}$$

which after simple manipulations becomes

$$i\, A^\rho{}_{00}\, d\mathbf{x}/dt + d^2\mathbf{x}/dt^2 = \tfrac{1}{2}\, \nabla\, h_{00}$$

In the limit where $\partial\Phi_k(x'')/\partial x''^\mu \approx 0$ for all k since it is driven by gravitation (as we will see later) we find the form of the Newtonian gravitational potential emerges:

$$d^2\mathbf{x}/dt^2 = \tfrac{1}{2}\, \nabla\, h_{00}$$

corresponding to

$$d^2\mathbf{x}/dt^2 = \tfrac{1}{2}\, \nabla\, \varphi(\mathbf{x})$$

of Newton.

5.8 Approximate Riemann-Christoffel Curvature Tensor and Einstein Dynamic Equations

The complex space-time Riemann-Christoffel curvature tensor (eq. 5.30) then is approximately (by eqs. 5.35-37)

$$R^\rho{}_{\mu\nu\sigma}(x'') \approx \partial\Gamma^\rho{}_{\mu\nu}(x')/\partial x'^\sigma - \partial\Gamma^\rho{}_{\mu\sigma}(x')\,/\partial x'^\nu + \partial\Gamma_V{}^\rho{}_{\mu\nu}(x'')/\partial x''^\sigma \;- \partial\Gamma_V{}^\rho{}_{\mu\sigma}(x'')/\partial x''^\nu +$$
$$+\, [\Gamma^\alpha{}_{\mu\nu}(x') + \Gamma_V{}^\alpha{}_{\mu\nu}(x'')]\, [\Gamma^\rho{}_{\sigma\alpha}(x') + \Gamma_V{}^\rho{}_{\sigma\alpha}(x'')] -$$

$$- [\Gamma^{a}{}_{\mu\sigma}(x') + \Gamma_V{}^{a}{}_{\mu\sigma}(x'')]\, [\Gamma^{\rho}{}_{\nu a}(x') + \Gamma_V{}^{\rho}{}_{\nu a}(x'')] \qquad (5.38)$$

using

$$\partial/\partial x''^{\sigma} = \partial x'^{\beta}/\partial x''^{\sigma}\, \partial/\partial x'^{\beta} \approx [U^{-1}(x'')^{\beta}{}_{\sigma} + x''^{\alpha}\, \partial U^{-1}(x'')^{\beta}{}_{\sigma}/\partial x''^{\beta}]\partial/\partial x'^{\beta} \approx \partial/\partial x'^{\sigma} \qquad (5.39)$$

to leading order. Rearranging terms we find

$$R^{\rho}{}_{\mu\nu\sigma}(x'') \approx R_{GR}{}^{\rho}{}_{\mu\nu\sigma}(x') + R_V{}^{\rho}{}_{\mu\nu\sigma}(x'') + R_{GRV}{}^{\rho}{}_{\mu\nu\sigma}(x') \qquad (5.40)$$

where

$$R_{GR}{}^{\rho}{}_{\mu\nu\sigma}(x') = \partial\Gamma^{\rho}{}_{\mu\nu}(x')/\partial x'^{\sigma} - \partial\Gamma^{\rho}{}_{\mu\sigma}(x')\, /\partial x'^{\nu} + \Gamma^{a}{}_{\mu\nu}(x')\Gamma^{\rho}{}_{\sigma a}(x') \; - \Gamma^{a}{}_{\mu\sigma}(x')\Gamma^{\rho}{}_{\nu a}(x') \qquad (5.41)$$

$$R_V{}^{\rho}{}_{\mu\nu\sigma}(x'') = \partial\Gamma_V{}^{\rho}{}_{\mu\nu}(x'')/\partial x''^{\sigma} - \partial\Gamma_V{}^{\rho}{}_{\mu\sigma}(x'')\, /\partial x''^{\nu} + \Gamma_V{}^{a}{}_{\mu\nu}(x'')\Gamma_V{}^{\rho}{}_{\sigma a}(x'') \; - $$
$$- \Gamma_V{}^{a}{}_{\mu\sigma}(x'')\Gamma_V{}^{\rho}{}_{\nu a}(x'') \qquad (5.42)$$

$$R_{GRV}{}^{\rho}{}_{\mu\nu\sigma}(x') = \Gamma^{a}{}_{\mu\nu}(x')\; \Gamma_V{}^{\rho}{}_{\sigma a}(x') + \Gamma_V{}^{a}{}_{\mu\nu}(x')\Gamma^{\rho}{}_{\sigma a}(x') - \Gamma^{a}{}_{\mu\sigma}(x')\Gamma_V{}^{\rho}{}_{\nu a}(x') - $$
$$- \Gamma_V{}^{a}{}_{\mu\sigma}(x')\Gamma^{\rho}{}_{\nu a}(x') \qquad (5.43)$$

letting $x'' \approx x'$ in eq. 5.43. The Ricci tensor is

$$R_{\mu\nu} = R^{a}{}_{\mu a\nu} \approx R_{GR}{}^{a}{}_{\mu a\nu}(x') + R_V{}^{a}{}_{\mu a\nu}(x'') + R_{GRV}{}^{a}{}_{\mu a\nu}(x') \qquad (5.44)$$

and the scalar curvature is

$$R(x'') = g^{\mu\nu}R_{\mu\nu}(x'') \approx g^{\mu\nu}[R_{GR}{}^{a}{}_{\mu a\nu}(x') + R_V{}^{a}{}_{\mu a\nu}(x'') + R_{GRV}{}^{a}{}_{\mu a\nu}(x')] \qquad (5.45)$$

where

$$g^{\mu\nu} \approx g^{\mu\nu}(x') \approx g^{\mu\nu}(x'')$$

The Einstein dynamic equation for this new Complex General Relativity is superficially familiar

$$R_{\mu\nu}(x'') - \tfrac{1}{2}\, g_{\mu\nu}\, R(x'') = -8\pi G\, T_{\mu\nu} \qquad (5.46)$$

Eq. 5.46 (unlike the real General Relativistic equation) consists of two parts, a real part and an imaginary part, which together jointly determine the dynamics of real-valued gravitation and the dynamics of the scalar fields. The imaginary part consists of sixteen equations for the sixteen scalar fields coupled to gravitation. The real part in our approximation describes gravitation as the standard real-valued equation with no coupling to the scalar fields. The scalar fields dynamic equations do have terms coupling them to gravity.

Thus we have a theory combining gravitation with scalar fields that is similar to the Extended Standard Model where fermions are coupled to gauge fields (and at large energies appear to become a GUT with a U(4)⊗U(4) symmetry as described in chapter 3.)

The real parts of the Ricci tensor and the scalar curvature *exactly*, in this approximation, yield the conventional *real-valued* Einstein dynamic equations

$$R_{GR\mu\nu}(x'') - \tfrac{1}{2}\, g_{\mu\nu}R_{GR}(x'') = -8\pi G\, T_{\mu\nu} \tag{5.47}$$

The approximate imaginary part of the Ricci tensor is

$$\begin{aligned} R_{I\mu\sigma} = \text{Im}\, R^{\alpha}{}_{\mu\alpha\sigma} \approx \text{Im}\, \{&\partial\Gamma_V{}^{\rho}{}_{\mu\rho}(x'')/\partial x''^{\sigma} - \partial\Gamma_V{}^{\rho}{}_{\mu\sigma}(x'')/\partial x''^{\rho} + \Gamma_V{}^{\alpha}{}_{\mu\rho}(x'')\Gamma_V{}^{\rho}{}_{\sigma\alpha}(x'') - \\ &- \Gamma_V{}^{\alpha}{}_{\mu\sigma}(x'')\Gamma_V{}^{\rho}{}_{\rho\alpha}(x'') + \Gamma_V{}^{\alpha}{}_{\mu\rho}(x'')\Gamma^{\rho}{}_{\sigma\alpha}(x') - \Gamma_V{}^{\alpha}{}_{\mu\sigma}(x'')\Gamma^{\rho}{}_{\rho\alpha}(x')\} \end{aligned} \tag{5.48}$$

The approximate imaginary part of the scalar curvature is

$$\begin{aligned} R_I = \text{Im}\, R = g^{\mu\sigma}R_{I\mu\sigma} = \text{Im}\{&\partial\Gamma_V{}^{\rho}{}_{\mu\rho}(x'')/\partial x''_{\mu} - \partial\Gamma_V{}^{\rho}{}_{\mu}{}^{\mu}(x'')/\partial x''^{\rho} + \Gamma_V{}^{\alpha}{}_{\mu\rho}(x'')\Gamma_V{}^{\rho\mu}{}_{\alpha}(x'') - \\ &- \Gamma_V{}^{\alpha}{}_{\mu}{}^{\mu}(x'')\Gamma_V{}^{\rho}{}_{\rho\alpha}(x'') + \Gamma_V{}^{\alpha}{}_{\mu\rho}(x'')\Gamma^{\rho\mu}{}_{\alpha}(x') - \Gamma_V{}^{\alpha}{}_{\mu}{}^{\mu}(x'')\Gamma^{\rho}{}_{\rho\alpha}(x')\} \end{aligned} \tag{5.49}$$

The imaginary part of the complex General Relativity Einstein equations

$$R_{I\mu\sigma} - \tfrac{1}{2}g_{\mu\nu}R_I(x'') = 0 \tag{5.50}$$

consists of a scalar particles' dynamic equation with interaction terms coupling to the real gravitational field and no direct coupling to the energy-momentum $T_{\mu\nu}$ (under the assumption $T_{\mu\nu}$ is real-valued although a tachyonic energy-momentum density would imply a complex $T_{\mu\nu}$).

Subsituting eq. 5.37 for $\Gamma_V{}^{\rho}{}_{\mu\rho}(x'')$ we find

$$R_{I\mu\sigma} \approx i\partial A^{\rho}{}_{\mu\rho}/\partial x''^{\sigma} - i\partial A^{\rho}{}_{\mu\sigma}/\partial x''^{\rho} - A^{\alpha}{}_{\mu\rho}A^{\rho}{}_{\sigma\alpha} + A^{\alpha}{}_{\mu\sigma}A^{\rho}{}_{\rho\alpha} + iA^{\alpha}{}_{\mu\rho}\Gamma^{\rho}{}_{\sigma\alpha} - iA^{\alpha}{}_{\mu\sigma}\Gamma^{\rho}{}_{\rho\alpha}$$

and

$$R_I \approx i\partial A^{\rho}{}_{\mu\rho}/\partial x''_{\mu} - i\partial A^{\rho\mu}{}_{\mu}/\partial x''^{\rho} - A^{\alpha}{}_{\mu\rho}A^{\rho\mu}{}_{\alpha} + A^{\alpha}{}_{\mu}{}^{\mu}A^{\rho}{}_{\rho\alpha} + iA^{\alpha}{}_{\mu\rho}\Gamma^{\rho}{}_{\mu\alpha} - iA^{\alpha}{}_{\mu}{}^{\mu}\Gamma^{\rho}{}_{\rho\alpha} \tag{5.51}$$

Substituting eq. 5.37a we find the *linear terms* in the imaginary part of the Einstein equation (eq. 5.50) are[23]

$$\tfrac{1}{2}i[\partial^2\varphi^\rho{}_\sigma/\partial x''^\rho\partial x''^\mu - \partial^2\varphi^\rho{}_\rho/\partial x''^\sigma\partial x''^\mu + \tfrac{1}{2}g_{\mu\sigma}(\Box\varphi^\rho{}_\rho - \partial^2\varphi^\rho{}_\beta/\partial x''^\rho\partial x''_\beta)] \qquad (5.52)$$

These terms plus the quadratic terms in eq. 5.50 constitute sixteen equations in the sixteen unknown fields $\Phi_k(x'')$. Since all terms in eq. 5.50 have only derivatives of $\Phi_k(x'')$, $\Phi_k(x'')$ is only specified up to an arbitrary constant c_k. We see

$$\Phi'_k(x'') = \Phi_k(x'') + c_k \qquad (5.53)$$

is also a solution of the imaginary Einstein equation for any choice of the 16 constants c_k.

5.9 The $\Phi_k(x'')$ Fields are the Higgs Fields

The freedom to shift the solutions $\Phi_k(x'')$ of the imaginary Einstein equation enables us to take these fields to be Higgs fields that can be used to give masses to gauge bosons and fermions in Extended Standard Model symmetry breaking. Usually, as in volume 1, we simply insert kinetic terms and potential terms in the Extended Standard Model lagrangian. However, we propose to, in a spirit of economy engendered by Ockham's Razor and Leibniz's MiniMax Principle, to take the imaginary Einstein lagrangian to be kinetic terms in The Theory of Everything lagrangian. We complete the Theory of Everything lagrangian by adding the Higgs potential energy terms to eq. 5.50 yielding the overall dynamic equations of the gravitational Higgs sector:

$$R_{I\mu\sigma} - \tfrac{1}{2}g_{\mu\nu}R_I(x'') = -8\pi i G_{Higgs}T_{Higgs\mu\sigma}(\Phi, V_{Higgs}(\Phi)) \qquad (5.54)$$

where G_{Higgs} is a constant, where the gravitational Higgs particle energy-momentum tensor $T_{Higgs\mu\sigma}$ is

$$T_{Higgs\mu\sigma}(\Phi, V_{Higgs}(\Phi)) = T_{HiggKinetics\mu\sigma}(\Phi) + \partial V_{Higgs}(\Phi)/\partial\Phi\ \partial\Phi/\partial x^\mu\ \partial\Phi(x)/\partial x^\sigma \qquad (5.54a)$$

and where $T_{HiggsKinetic\mu\sigma}(\Phi)$ is the kinetic part of the energy-momentum tensor for the gravitational Higgs fields.

The shift of gravitational Higgs fields to their vacuum expectation values is allowed due to derivatives appearing in all left side terms of eq. 5.54.

Consequently Gravitational Higgs particles can have a vacuum expectation value Φ_{k0} such that the Higgs field is a sum of a vacuum expectation value and a field $\varphi_k(x)$ dependent on x:

---

[23] Since the quadratic terms do not add to the significance of our discussion we defer their discussion, and the discussion of solutions, to a subsequent time.

$$\Phi'_k(x) = \varphi_k(x) + \Phi_{k0} \qquad (5.55)$$

5.10 Appearance of Gravitational Higgs Fields in Fermion Dynamic Equations

The fermion equations have Higgs terms that become masses (plus Higgs fields) after spontaneous breakdown. The introduction of a gravitational Higgs term in the fermion dynamic equations must be as a scalar term:

$$g^{\mu}{}_{\sigma}\varphi^{\sigma}{}_{\mu} = \sum_k g_k\Phi_k(x'')g^{\mu}{}_{\sigma}[\tau_k]^{\sigma}{}_{\mu} = \sum_k g_k\Phi_k(x'')[\tau_k]^{\sigma}{}_{\sigma}$$

The τ_k matrices are the generators of U(4). They have been transformed using the tetrad formalism (eq. 5.15a)

$$[\tau_k]^{\mu}{}_{\upsilon} = w^{\mu}{}_{a}(x)\,[\tau_k]^{a}{}_{b}\,v^{b}{}_{\upsilon}(x)$$

and so we find the trace of the generators

$$[\tau_k]^{\sigma}{}_{\sigma} = [\tau_k]^{a}{}_{a}$$

Since there are four diagonal generators in the 4-dimensional U(4) representation we find the gravitational Higgs term in each fermion dynamic equation is

$$g^{\mu}{}_{\sigma}\varphi^{\sigma}{}_{\mu} = \sum_k g_k\Phi_k(x'')$$

where the sum over k is over the k values of the four diagonal τ matrices. Inserting eq. 5.55 we find the gravitational Higgs term in each fermion dynamic equation is:

$$g^{\mu}{}_{\sigma}\varphi^{\sigma}{}_{\mu} = \sum_k g_k\,[\Phi_{k0} + \varphi_k(x)] \qquad (5.55a)$$

yielding the mass term Φ_0

$$\sum_k g_k\Phi_{k0} = \Phi_{Grav0} \qquad (5.55b)$$

which contains four gravitational Higgs particles vacuum expectation values.[24] Thus twelve of the gravitational Higgs particles need not have non-zero vacuum expectation values although they have dynamic significance due to the $\varphi_k(x)$ fields' dynamic equations.

24 The constants g_k can be set equal to a constant g_{grav}.

5.11 Fermion Masses Contributions from Gravitational Higgs Particles

In chapter 2 and in volume I we discussed the fermion mass contributions from the vacuum expectation values of ElectroWeak and Dark ElectroWeak Higgs particles, and from vacuuu expectation values of Generation group Higgs particles (chapter 2 and chapter 16 volume I). In this section we will complete the picture for fermion mass contributions by adding contributions from the Gravitational Higgs particles vacuuu expectation values that we described in this chapter.

Introducing Gravitational Higgs particles the lagrangian mass terms for the four normal and four Dark fermion species (of the above discussion, eq. 16.48, and volume I) become

$$\mathcal{L}^{Higgs}{}_{FermionMasses} = \Sigma_{\alpha,\beta,k}\, \bar{\psi}_{kL\alpha}\, \eta m_{EW_{k\alpha\beta}} \psi_{kR\beta} + \Sigma_{\alpha,\beta}\, \bar{\psi}_{D_{kL\alpha}}\, \eta_D m_{DEW_{k\alpha\beta}} \psi_{D_{kR\beta}} +$$

ElectroWeak

$$+ \Sigma_{\alpha,\beta} \{\bar{\psi}_{UuL\alpha}\, \eta_{Uu} m_{Uu\alpha\beta} \psi_{UuR\beta} + \bar{\psi}_{UdL\alpha}\, \eta_{Ud} m_{Ud\alpha\beta} \psi_{UdR\beta}\} +$$
$$+ \Sigma_{\alpha,\beta} \{\bar{\psi}_{DUuL\alpha} \eta_{DUu} m_{DUu\alpha\beta} \psi_{DUuR\beta} + \bar{\psi}_{DUdL\alpha} \eta_{DUd} m_{DUd\alpha\beta} \psi_{DUdR\beta}\} +$$

Generation Group U

$$+ \Sigma_{\alpha,\beta,k}\, \bar{\psi}_{G_{kL\alpha}}\, \eta_G m_{G_{k\alpha\beta}} \psi_{G_{kR\beta}} + \Sigma_{\alpha,\beta,k}\, \bar{\psi}_{DGkL\alpha}\, \eta_{DG} m_{DGk\alpha\beta} \psi_{DGkR\beta} +$$

Gravitational Reality Group

$$+ \text{c.c.} \qquad (5.56)$$

where the subscripts EW, u, d, D, U and G label ElectroWeak origin, up-type quark species, down-type quark species, Dark type, Generation group origin, and Gravitational origin respectively. The fields labeled η (with subscripts) are Higgs fields that have non-zero vacuum expectation values. The index k = 1, … 4 labels species. The matrices labeled m (with subscripts) are the complex constant mass matrices of species. The indices α, β = 1, …, 4 label rows and columns.

5.12 The Full Extended Standard Model Fermion Mass Matrices

Combining the terms in eq. 5.56 for each species we obtain their total mass matrices below which can then be diagonalized to obtain the masses of the fermions within each species.

Charged Lepton Species Total Mass Matrix

$$m_{etot} = m_{EWe} + m_{Ge} \qquad (5.57)$$

Neutral Lepton Species Mass Matrix

$$m_{\upsilon tot} = m_{EW\upsilon} + m_{G\upsilon} \qquad (5.58)$$

Up-Type Quark Species Mass Matrix

$$m_{utot} = m_{EWu} + m_{Uu} + m_{Gu} \qquad (5.59)$$

Down-Type Quark Species Mass Matrix

$$m_{dtot} = m_{EWd} + m_{Ud} + m_{Gd} \quad (5.60)$$

Dark Charged Lepton Species Total Mass Matrix

$$m_{Detot} = m_{DEWe} + m_{DGe} \quad (5.61)$$

Dark Neutral Lepton Species Mass Matrix

$$m_{D\upsilon tot} = m_{DEW\upsilon} + m_{DG\upsilon} \quad (5.62)$$

Dark Up-Type Quark Species Mass Matrix

$$m_{Dutot} = m_{DEWu} + m_{DUu} + m_{DGu} \quad (5.63)$$

Dark Down-Type Quark Species Mass Matrix

$$m_{Ddtot} = m_{DEWd} + m_{DUd} + m_{DGd} \quad (5.64)$$

We now note that the preceding formal development yields $m_{Ge} = m_{G\upsilon} = m_{Gu} = m_{Gd} = m_{DGe} = m_{DG\upsilon} = m_{DG\upsilon} = m_{DGu} = m_{DGd} = m_G$. The gravitational contribution to all fermions of all species is the same.

Moreover, the gravitational contribution to each fermion mass sets the scale for all fermion masses (and secondarily of massive gauge bosons' masses) yielding the "principle" of Newton, Einstein and others that *inertial mass equals to gravitational mass*. (See chapter 6.)

The generation group contributions, in the spontaneous breakdown that we described, appear only in quark and Dark quark mass matrices providing, possibly, a reason why quark masses are so much larger than lepton masses.

The mass matrices above can each be diagonalized in a manner similar to that of eqs. 16.50 and 16.51 in chapter 2 and volume I.

Appendix 5-A. Features of Complex General Relativity

EXTRACTED FROM BLAHA(2004). This appendix may be skipped, at first reading, since it is not important for the remaining chapters.

5-A.1 Introduction

In this apendix we will discuss an extension of Riemannian geometry for real space-time to the case of *complex four-dimensional space-time*[25] without the use of the Reality group generators. The reader is assumed to be conversant with the conventional forms of Riemannian geometry, tensor analysis and general relativity. (There are many excellent texts on these subjects.)

While it would be easy to develop the discussion for an arbitrary number of dimensions we will assume a four dimensional space-time in our development. The generalization to other dimensions is direct.

5-A.2 The Fundamental Quadratic Differential Form

Riemannian geometry begins with the study of the fundamental quadratic form:

$$d\tau^2 = g_{\mu\nu}dX^\mu dX^\nu \qquad (5\text{-A.2.1})$$

where $d\tau^2$ is the square of $d\tau$. If $d\tau^2$ is positive then $d\tau$ is the *proper time*. If $d\tau^2$ is negative then $|d\tau|$ is the *proper distance*. In either case $|d\tau|$ is a measure of the interval between two nearby points $X = (X^0, X^1, X^2, X^3)$ and $X + dX = (X^0 + dX^0, X^1 + dX^1, X^2 + dX^2, X^3 + dX^3)$. The space-time points in Riemannian geometry are real numbers and the metric $g_{\mu\nu}$ consists of sixteen real (but not independent) quantities. In a flat space-time the metric takes the simple form: $g_{00} = 1$; $g_{ij} = -\delta_{ij}$ for i, j = 1, 2, 3; and $g_{\mu\nu} = 0$ if $\mu \neq \nu$. Thus $d\tau^2$ is positive and $d\tau$ is a real number if the interval is time-like; and $d\tau^2$ is negative and $d\tau$ is an imaginary number if the interval is space-like.

If we were to plot values of the $d\tau$ complex plane we would see its values lie on the real and imaginary axes as shown in Fig. 5-A.2.1. Eddington[26] commented on this point observing, "This world-geometry has a property unlike that of Euclidean geometry in that the interval

---

[25] R. Penrose and collaborators have developed an 8-dimensional (4 complex dimensions) space-time in their twistor theory. This development differs in fundamental ways from the present discussion. See R. Penrose and M. A. H. MacCallum, Physics Reports **6**, 241 (1972) and references therein. Other complex 5-dimensional theories are described in Streater (2000) and Charon (1988). These also differ significantly from our development.

[26] Eddington (1995) p. 150.

between two real events may be real or imaginary. The necessity for a physical distinction, corresponding to the mathematical distinction between real and imaginary intervals, introduces us to the separation of the four-dimensional order into time and space."

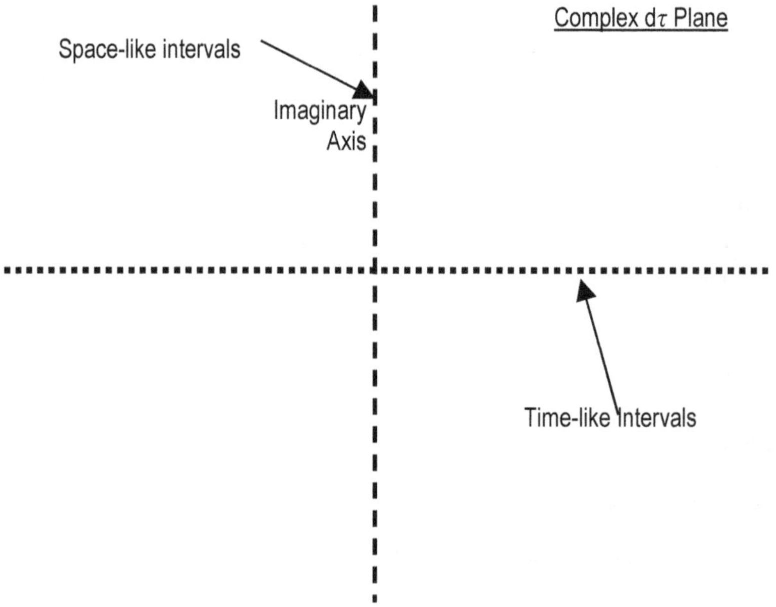

Figure 5-A.2.1. Plot of the complex dτ plane for real-valued intervals. Space-like intervals are purely imaginary and thus lie on the y-axis. Time-like intervals are purely real and lie on the x-axis.

No special significance is attributed to the fact that dτ is imaginary for space-like distances. The numeric value of dτ (modulo i) is taken to be the space-like distance between points (or events). Thus the concept of proper time or interval readily encompasses real or imaginary values.

5-A.3 Complex Space-Time

We now wish to consider the possibility of complex values of dτ - not just purely real or imaginary values. Naturally this requires a complex space-time manifold – not the real space-time manifold of everyday experience.

Suppose we consider a complex four-dimensional space in which the four-vectors have complex values and $g_{\mu\nu}$ is correspondingly complex. If we choose a Galilean coordinate system (rectangular coordinates) then we can define an interval with the metric form:

$$d\tau^2 = g_{\mu\nu}dX^\mu dX^\nu \qquad (5\text{-A.3.1})$$

Note $g_{\mu\nu} = g_{\nu\mu}$ is symmetric in complex space-time.

The discerning reader will also note that, for good reason, we have used $dX^\mu dX^\nu$ in eq. 5-A.3.1 with the result that $d\tau^2$ is complex in general. (The conventional choice[27] is $g_{\mu\nu}dX^\mu d\bar{X}^\nu$ where $d\bar{X}^\nu$ is the complex conjugate of dX^ν and where $g_{\mu\nu} = g_{\nu\mu}{}^*$. This choice guarantees the interval (distance) is a real number $d\tau^{2*} = d\tau^2$.) The reasons for the definition eq. 5-A.3.1 are:

1. Practical – It will be seen that this choice (eq. 5-A.3.1) leads to a theory of quantum gravity that is finite whereas choosing the conventional form $g_{\mu\nu}dX^\mu d\bar{X}^\nu$ leads to a divergent, non-renormalizable theory.

2. Esthetic – The definition of interval given by eq. 5-A.3.1 enables $g_{\mu\nu}$ to be a holomorphic (analytic) function of the coordinates – a feature that survives a change of coordinate system. Thus we obtain the benefits of analyticity.

3. Mathematical – The definition of eq. 5-A.3.1 supports an almost complex map from an eight-dimensional space to our four-dimensional complex space.[28]

4. Rational – The definition of eq. 5-A.3.1 becomes equivalent to the standard definition of interval in the purely real subspace that constitutes a conventional real Riemannian space. Thus we can determine the analytic extension without guesswork from the real space-time limit.

5-A.4 Complex Coordinate Transformations

We now consider complex coordinate transformations under which the complex interval $d\tau^2$ (that is specified by eq. 5-A.3.1) is invariant. If X'^μ denotes an alternate complex coordinate system that is related to X^μ by continuous, holomorphic functions f^μ:

$$X^\mu = f^\mu(X'^0, X'^1, X'^2, X'^3) \qquad (5\text{-A.4.1})$$

for $\mu = 0, 1, 2, 3$ then dX^μ are vectors:

$$dX^\mu = dX'^\nu\, \partial f^\mu(X'^0, X'^1, X'^2, X'^3)/\partial X'^\nu \qquad (5\text{-A.4.2})$$

or, more simply,

[27] Lang (1987) discusses length functions and distance measures in complex spaces in some detail. Our choice of interval differs from the cases that he considers. Yet we will see it offers a physically acceptable interpretation.

[28] A. Newlander and L. Nirenberg, *Complex Analytic Coordinates in Almost Complex Manifolds*, Annals of Mathematics **65**, 391 (1957); E. Calabi, *Construction and Properties of Some 5-Dimensional Almost Complex Manifolds*, Transactions of the American Mathematical Society **87**, 407 (1958).

$$dX^{\mu} = dX'^{\nu}\, \partial X^{\mu}/\partial X'^{\nu} \qquad (5\text{-A.4.3})$$

by the rules of partial differentiation. (In this book we define holomorphic in a conventional to mean analytic in *some* domain. Thus we will say $g_{\mu\nu}$, as we define it, is an holomorphic function of X^{μ}. Nevertheless $g_{\mu\nu}$ can have singularities, branch points, etc.) Since the components of each X^{μ} vector is a complex variable, we can define the partial derivative of any function, or tensor, with respect to a coordinate X^{ν} to be:

$$\partial f(X)/\partial X^{\nu} = \lim_{w^{\nu} \to 0} [f(X^{\nu} + w^{\nu}) - f(X^{\nu})]/w^{\nu} \qquad (5\text{-A.4.4})$$

with no implied summation over ν and the other coordinates X^{μ} held fixed where $\mu \neq \nu$. The only non-zero component of the vector w^{μ} is the νth component w^{ν} which is an arbitrary complex quantity that approaches zero in the limit. From a geometric viewpoint, the point $X^{\nu} + w^{\nu}$ can approach X^{ν} along any arbitrary curve in the complex X^{ν} plane in eq. 5-A.4.4. The limit in eq. 5-A.4.4 must exist and be independent of the path along which $X^{\nu} + w^{\nu}$ approaches X^{ν}. The quantity $f(X^{\mu})$ is *differentiable* at the point X^{μ} if all first order partial derivatives with respect to all components of X^{μ} exist at the point X^{μ}.

Since the definition of a complex partial derivative is formally identical to the partial derivative with respect to a real variable we can make the following observation.

Observation: *All the formal rules of complex partial differentiation are the same as those of real partial differentiation. Thus the differentiation of sums, products, quotients and so on are formally the same as in the real variable case. In particular, a complex partial derivative will coincide with the corresponding real partial derivative on the real axis (i.e. for X^{ν} real).*

Thus we are saved the labor of recalculating the equations of Riemannian geometry for the complex case.

Under complex coordinate transformations the metric tensor transforms as a covariant second rank tensor:

$$g_{\mu\nu} = g'_{\alpha\beta}\, \partial X'^{\alpha}/\partial X^{\nu}\, \partial X'^{\beta}/\partial X^{\nu} \qquad (5\text{-A.4.5})$$

$d\tau^2$ is invariant under complex coordinate transformations:

$$d\tau^2 = g_{\mu\nu} dX^{\mu} dX^{\nu} = d\tau'^2 = g'_{\mu\nu} dX'^{\mu} dX'^{\nu} \qquad (5\text{-A.4.6})$$

The inverse of $g_{\mu\nu}$, denoted $g^{\mu\nu}$, satisfies

$$g_{\alpha\nu}\, g^{\nu\beta} = \delta_{\alpha}^{\ \beta} \qquad (5\text{-A.4.7})$$

where $\delta_\alpha^{\ \beta}$ is the Kronecker delta function ($\delta_\alpha^{\ \beta} = 1$ if $\alpha = \beta$ and is zero otherwise), and transforms as a contravariant tensor under a complex coordinate transformation:

$$g'^{\mu\nu} = g^{\alpha\beta}\, \partial X'^{\mu}/\partial X^{\alpha}\ \partial X'^{\nu}/\partial X^{\beta} \qquad (5\text{-A.4.8})$$

5-A.5 Analyticity of the Complex Coordinates

Complex space-time must meet two physical requirements in order to be consistent with general relativity when we restrict the coordinates to its real subspace:

Requirement 1. *In the real subspace sector* the Principle of Equivalence must apply so that the motion of a particle, moving under the influence of gravitational forces only, can be described as straight line motion in a free falling coordinate system x^μ:

$$d^2x^\mu/d\tau^2 = 0 \qquad (5\text{-A.5.1})$$

where

$$d\tau^2 = \eta_{\mu\nu}dx^\mu dx^\nu \qquad (5\text{-A.5.2})$$

with $\eta_{00} = 1$; $\eta_{ij} = -\delta_{ij}$ for i, j = 1, 2, 3; and $\eta_{\mu\nu} = 0$ if $\mu \neq \nu$.

Requirement 2. Complex coordinate transformations such as eq. 5-A.4.1 must be continuous and holomorphic.

Consequently, the metric tensor for an arbitrary coordinate system X^μ can be written as:

$$g_{\mu\nu}(X) = \eta_{\alpha\beta}\, \partial x^{\alpha}/\partial X^{\nu}\ \partial x^{\beta}/\partial X^{\nu} \qquad (5\text{-A.5.3})$$

where $g_{\mu\nu}(X)$ is an holomorphic function of X^μ. Therefore

$$\partial g_{\mu\nu}/\partial \overline{X}^{\rho} = 0 \qquad (5\text{-A.5.4})$$

where $\overline{X}^\rho$ is the complex conjugate of X^ρ. The analyticity in X^ρ can also be expressed equivalently as a set of Cauchy-Riemann equations.

If we express each coordinate X^ρ in terms of its real and imaginary parts

$$X^\rho = {}_{r}X^\rho + i\ {}_{i}X^\rho \qquad (5\text{-A.5.5})$$

and the metric tensor in terms of its real and imaginary parts

$$g_{\mu\nu} = {}_{r}g_{\mu\nu} + i\ {}_{i}g_{\mu\nu} \qquad (5\text{-A.5.6})$$

then the Cauchy-Riemann equations are

$$\partial_r g_{\mu\nu}/\partial_r X^\rho = \partial_i g_{\mu\nu}/\partial_i X^\rho \qquad (5\text{-A.5.7})$$

and

$$\partial_r g_{\mu\nu}/\partial_i X^\rho = -\,\partial_i g_{\mu\nu}/\partial_r X^\rho \qquad (5\text{-A.5.8})$$

for all μ, ν and ρ.

The analyticity of $g_{\mu\nu}(X)$ in X^ρ will ultimately result in a finite theory of Quantum Gravity.[29]

The tensor analysis of complex coordinate transformations of the type that we are considering is very similar to the tensor analysis for real coordinate transformations since the manipulations are largely algebraic in nature. As a result the expressions and equations look the same[30] although their interpretation and consequences are different.

The complex partial derivative $\partial/\partial X^\nu$ of a function of X with respect to X^ν can be formally expressed in terms of the real and imaginary parts of X^ν by:

$$\partial f(X)/\partial X^\rho = \tfrac{1}{2}\,[\partial f(X)/\partial_r X^\rho - i\,\partial f(X)/\partial_i X^\rho] \qquad (5\text{-A.5.9})$$

and the complex partial derivative of a function $h(X, \bar{X})$ with respect to the complex conjugate of X^ν, namely $\partial/\partial\bar{X}^\nu$, can be formally expressed by:

$$\partial h/\partial \bar{X}^\rho = \tfrac{1}{2}\,[\partial h/\partial_r X^\rho + i\,\partial h/\partial_i X^\rho] \qquad (5\text{-A.5.10})$$

We note that if h is only a function of X (and not $\bar{X}$) $h = h(X)$, then

$$\partial h(X)/\partial \bar{X}^\rho = 0 \qquad (5\text{-A.5.11})$$

Eq. 5-A.5.11 is equivalent to the Cauchy-Riemann equations for an holomorphic function of the complex variables X^ρ.

---

[29] S. Blaha, *A Finite Unified Quantum Field Theory of the Elementary Particle Standard Model and Quantum Gravity Based on New Quantum Dimensions™ and a New Paradigm in the Calculus of Variations* (Pingree-Hill Publishing, Auburn, NH, 2003).

[30] See, for example, C. W. Misner, K. S. Thorne and J. A. Wheeler, *Gravitation*, (W. H. Freeman, San Francisco, 1973); S. Weinberg, *Gravitation and Cosmology* (John Wiley & Sons, New York, 1972); H. Weyl, tr. H. L. Brose, *Space, Time, Matter* (Dover Publications, New York, 1952); A. Eddington, *Space, Time & Gravitation* (Cambridge University Press, Cambridge, 1920 and 1995).

5-A.5.1 Motion of a Particle

Eqs. 5-A.5.1 and 5-A.5.2 describe the motion of a particle moving only under gravitational forces in a free falling coordinate system x^μ. If we implement a holomorphic transformation to a new coordinate system:

$$x^\mu = f^\mu(X) \qquad (5\text{-A.5.1.1})$$

with metric tensor:

$$g_{\mu\nu}(X) = \eta_{\alpha\beta}\, \partial x^\alpha/\partial X^\nu\, \partial x^\beta/\partial X^\nu \qquad (5\text{-A.5.1.2})$$

then the motion of the particle in the new coordinate system is described by the equation of motion:

$$d^2X^\mu/d\tau^2 + \Gamma^\mu{}_{\lambda\sigma}\, dX^\lambda/d\tau\, dX^\sigma/d\tau \;= 0 \qquad (5\text{-A.5.1.3})$$

where

$$d\tau^2 = g_{\mu\nu} dX^\mu dX^\nu \qquad (5\text{-A.5.1.4})$$

with the affine connection given by

$$\Gamma^\sigma{}_{\lambda\mu} = \partial X^\sigma/\partial x^\rho\; \partial^2 x^\rho/\partial X^\lambda \partial X^\mu \qquad (5\text{-A.5.1.5})$$

The affine connection satisfies the symmetry condition

$$\Gamma^\sigma{}_{\lambda\mu} = \Gamma^\sigma{}_{\mu\lambda} \qquad (5\text{-A.5.1.6})$$

Note X^σ, $g_{\mu\nu}(X)$, and $\Gamma^\sigma{}_{\lambda\mu}$ are all complex quantities.

5-A.5.2 The Complex Affine Connection

Due to the analyticity of $g_{\mu\nu}(X)$, which is maintained under complex, holomorphic coordinate transformations, and due to the algebraic nature of the mathematical manipulations, the complex affine connection has the same form as the affine connection in real Riemannian space-times. It can be expressed in terms of the metric tensor as:

$$\Gamma^\sigma{}_{\lambda\mu} = \tfrac{1}{2}\, g^{\nu\sigma}\{\partial g_{\mu\nu}/\partial X^\lambda + \partial g_{\lambda\nu}/\partial X^\mu - \partial g_{\lambda\mu}/\partial X^\nu\} \qquad (5\text{-A.5.2.1})$$

The proof is algebraically identical to the real case.

5-A.5.3 Complex Covariant Derivatives

As in the case of real space-time, the derivative of a complex tensor is not necessarily another tensor. A simple example is the case of a contravariant vector Z^μ, which transforms as:

$$Z'^{\nu} = Z^{\beta}\, \partial X'^{\nu}/\partial X^{\beta} \tag{5-A.5.3.1}$$

The derivative

$$\partial Z'^{\nu}/\partial X'^{\rho}$$

is not a tensor. A covariant derivative can be defined using the affine connection just as in the case of real space-time. The covariant derivative of a contravariant vector Z^ν is

$$Z^{\nu}_{\ ;\mu} = \partial Z^{\nu}/\partial X^{\mu} + \Gamma^{\nu}_{\ \mu\lambda}\, Z^{\lambda} \tag{5-A.5.3.2}$$

$Z^{\nu}_{\ ;\mu}$ is a mixed tensor. Similarly the covariant derivative of a covariant vector:

$$Z_{\nu;\mu} = \partial Z_{\nu}/\partial X^{\mu} - \Gamma^{\lambda}_{\ \nu\mu}\, Z_{\lambda} \tag{5-A.5.3.3}$$

is a covariant tensor.

The forms of the covariant derivative of covariant, contravariant and mixed tensors, and tensor densities, are the same in form as their real space-time counterparts as eqs. 5-A.5.3.2 and 5-A.5.3.3 show.

An important special case of the above is the covariant derivative of the metric tensor:

$$g_{\mu\nu;\sigma} = \partial g_{\mu\nu}/\partial X^{\sigma} - \Gamma^{\lambda}_{\ \sigma\mu} g_{\lambda\nu} - \Gamma^{\lambda}_{\ \sigma\nu} g_{\lambda\mu} \tag{5-A.5.3.4}$$

Since $\partial g_{\mu\nu}(X)/\partial X^{\rho}$, and $\Gamma^{\sigma}_{\ \lambda\mu}$ are zero in a locally inertial reference frame and since a tensor that has the value zero in one coordinate system will have the value zero in all coordinate systems, we see that

$$g_{\mu\nu;\sigma} = 0 \tag{5-A.5.3.5}$$

As a result the operations of raising and/or lowering indices commutes with covariant differentiation:

$$(g_{\mu\nu} Z^{\nu})_{;\sigma} = g_{\mu\nu} Z^{\nu}_{\ ;\sigma} \tag{5-A.5.3.6}$$

In addition, the expressions for covariant derivatives in real space-time have the same form in complex space-time. For example,

1. The covariant derivative of a scalar quantity:

$$S_{;\rho} = \partial S/\partial X^\rho \tag{5-A.5.3.7}$$

2. The covariant divergence of a vector:

$$V^\rho{}_{;\rho} = g^{-1/2}\, \partial(g^{1/2}V^\rho)/\partial X^\rho \tag{5-A.5.3.8}$$

using

$$\Gamma^\sigma{}_{\sigma\mu} = g^{-1/2}\, \partial(g^{1/2})/\partial X^\mu \tag{5-A.5.3.9}$$

5-A.5.4 The Complex Curvature Tensor

The complex space-time Riemann-Christoffel curvature tensor has the same form as its real space-time equivalent:

$$R^\rho{}_{\mu\nu\sigma} \equiv \partial\Gamma^\rho{}_{\mu\nu}/\partial X^\sigma - \partial\Gamma^\rho{}_{\mu\sigma}/\partial X^\nu + \Gamma^\alpha{}_{\mu\nu}\,\Gamma^\rho{}_{\sigma\alpha} - \Gamma^\alpha{}_{\mu\sigma}\,\Gamma^\rho{}_{\nu\alpha} \tag{5-A.5.4.1}$$

Some related quantities are:

1. The Ricci tensor:

$$R_{\mu\nu} = R^\alpha{}_{\mu\alpha\nu} \tag{5-A.5.4.2}$$

2. The curvature scalar:

$$R = g^{\mu\nu}R_{\mu\nu} \tag{5-A.5.4.3}$$

3. The covariant form of the curvature tensor:

$$R_{\rho\mu\nu\sigma} = g_{\rho\alpha}\, R^\alpha{}_{\mu\nu\sigma} \tag{5-A.5.4.4}$$

which also can be written

$$R_{\rho\mu\nu\sigma} = \tfrac{1}{2}\,[\partial^2 g_{\rho\nu}/\partial X^\sigma\partial X^\mu - \partial^2 g_{\mu\nu}/\partial X^\sigma\partial X^\rho - \partial^2 g_{\rho\sigma}/\partial X^\nu\partial X^\mu + \partial^2 g_{\mu\sigma}/\partial X^\nu\partial X^\rho]$$

$$+ g_{\alpha\beta}[\Gamma^\alpha{}_{\nu\rho}\,\Gamma^\beta{}_{\mu\sigma} - \Gamma^\alpha{}_{\sigma\rho}\,\Gamma^\beta{}_{\mu\nu}] \tag{5-A.5.4.5}$$

Because of *the exact correspondence between the algebraic properties and differential properties of the real space-time and our complex space-time curvature tensors*, the properties and identities of the complex space-time curvature tensor are:

1. The Algebraic Symmetry Properties:

$$R_{\rho\mu\nu\sigma} = R_{\nu\sigma\rho\mu} \qquad (5\text{-A.5.4.6})$$

$$R_{\rho\mu\nu\sigma} = R_{\mu\rho\sigma\nu} = -R_{\rho\mu\sigma\nu} = -R_{\mu\rho\nu\sigma} \qquad (5\text{-A.5.4.7})$$

$$R_{\rho\mu\nu\sigma} + R_{\rho\sigma\mu\nu} + R_{\rho\nu\sigma\mu} = 0 \qquad (5\text{-A.5.4.8})$$

$$R_{\mu\nu} = R_{\nu\mu} \qquad (5\text{-A.5.4.9})$$

2. The Bianchi Identities:

$$R_{\rho\mu\nu\sigma\,;\,\lambda} + R_{\rho\mu\lambda\nu\,;\,\sigma} + R_{\rho\mu\sigma\lambda;\nu} = 0 \qquad (5\text{-A.5.4.10})$$

3. The covariant derivative:

$$(R^{\mu\nu} - \tfrac{1}{2}\, g^{\mu\nu}R)_{;\,\mu} = 0 \qquad (5\text{-A.5.4.11})$$

5-A.5.5 The Interval Between Events in Complex Space-time

Much of the physical interpretation of the events and structure of a space-time depends on an understanding of the space-time's notion of an interval. As we saw earlier in the discussion of the motion of a particle in section 5-A.5.1 the proper time interval $d\tau$ parameterizes motion.

However the intervals in complex space-time are not only time-like or space-like. In general, they are a complex combination of both time-like and space-like parts. In this section we will develop a characterization of complex intervals for later use in the development of a complex formulation of general relativity.

We will begin by expanding the various quantities in terms of their real and imaginary parts. If we expand eq. 5-A.3.1 using

$$g_{\mu\nu} = {}_{r}g_{\mu\nu} + i\ {}_{i}g_{\mu\nu} \qquad (5\text{-A.5.5.1})$$

and

$$dX_{\mu} = {}_{r}dX_{\mu} + i\ {}_{i}dX_{\mu} \qquad (5\text{-A.5.5.2})$$

we find

$$d\tau^2 = {}_{r}g_{\mu\nu}({}_{r}dX^{\mu}\,{}_{r}dX^{\nu} - {}_{i}dX^{\mu}\,{}_{i}dX^{\nu}) + 2i\ {}_{r}g_{\mu\nu}\,{}_{r}dX^{\mu}\,{}_{i}dX^{\nu} + $$
$$+\ i\ {}_{i}g_{\mu\nu}({}_{r}dX^{\mu}\,{}_{r}dX^{\nu} - {}_{i}dX^{\mu}\,{}_{i}dX^{\nu}) - 2\ {}_{i}g_{\mu\nu}\,{}_{r}dX^{\mu}\,{}_{i}dX^{\nu} \qquad (5\text{-A.5.5.3})$$

where the real and imaginary parts of the metric are assumed to be symmetric in their indices:

$$_{\mathrm{r}}g_{\mu\nu} = {}_{\mathrm{r}}g_{\nu\mu} \tag{5-A.5.5.4}$$

and

$$_{\mathrm{i}}g_{\mu\nu} = {}_{\mathrm{i}}g_{\nu\mu} \tag{5-A.5.5.5}$$

An examination of the form of the proper time intervals in eq. 5-A.5.5.3 shows that $d\tau^2$, and $d\tau$, are both complex numbers in general.

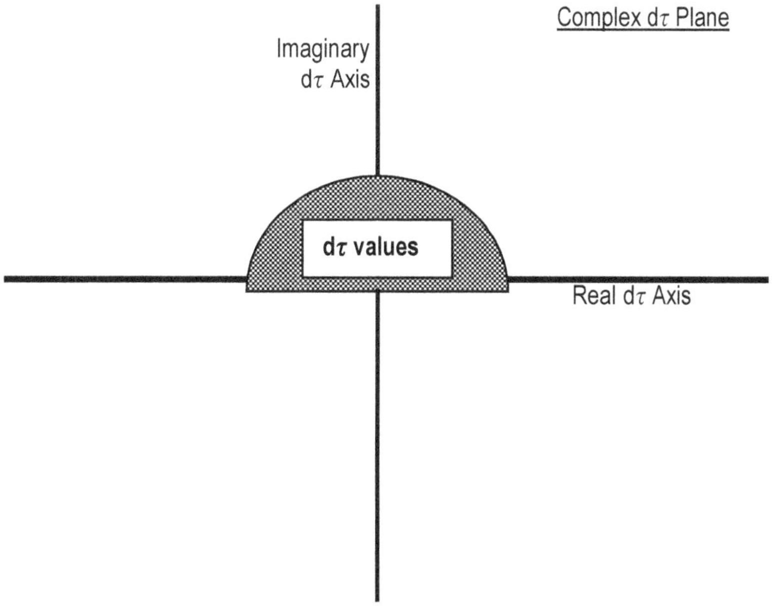

Figure 5-A.5.5.1. Schematic plot of the complex $d\tau$ plane. In real space-time space-like intervals are purely imaginary and thus lie on the y-axis and time-like intervals are purely real and lie on the x-axis.

The values of $d\tau$ are complex infinitesimal values in the upper complex $d\tau$ plane as shown in Fig. 5-A.5.5.1. If we assume the real and imaginary parts of dX^μ are infinitesimal then the allowed values of $d\tau$ cluster in something like a roughly infinitesimal "semicircle" centered at 0. A complex interval $d\tau$ can be viewed as partly time-like and partly space-like.

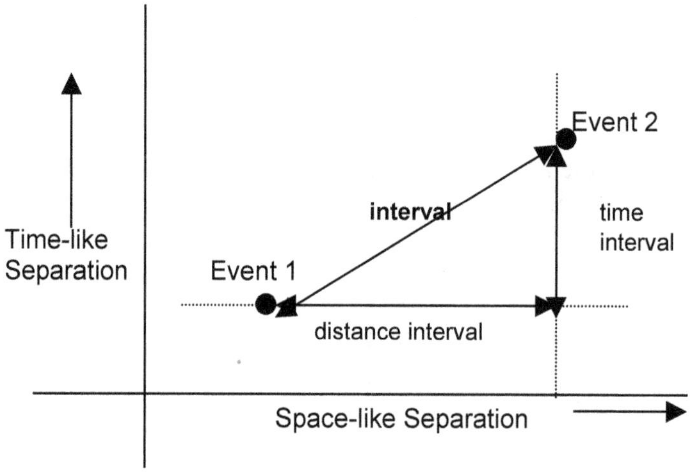

Figure 5-A.5.5.2. The relation between two events that have both a time-like and space-like separation.

In real space-time the interval between events, a physical quantity, is *assumed* to be the square root of the fundamental quadratic form $d\tau^2$ which is a real, positive or negative or null-valued, quantity. When $d\tau^2$ is space-like (negative) the inconvenient factor of $i = \sqrt{-1}$ is discarded.

In complex space-time $d\tau^2$ is complex, and its square root is complex as well, thus creating a question of physical interpretation. Therefore we will define the *physical interval* ds between events to be:

$$ds = \sqrt{[(\text{Re } d\tau)^2 + (\text{Im } d\tau)^2]} \qquad (5\text{-A.5.5.6})$$

In the limit that the metric and the intervals are purely real ds becomes equivalent to the conventional definition of space-like and time-like intervals.

The interval ds can also be represented in the following ways:

$$ds = \sqrt{[d\tau d\tau^*]} \qquad (5\text{-A.5.5.7})$$

and

$$ds = \sqrt[4]{[(\text{Re } d\tau^2)^2 + (\text{Im } d\tau^{2*})^2]} \qquad (5\text{-A.5.5.8})$$

$$= \sqrt[4]{[d\tau^2 d\tau^{2*}]} \qquad (5\text{-A.5.5.9})$$

We note that if $d\tau^2$ is invariant under holomorphic coordinate transformations then $d\tau^{2*}$ is invariant under "anti-holomorphic" coordinate transformations (the complex conjugate of the holomorphic transformations). Thus the physical interval ds is invariant under these coordinate transformations.

5-A.5.6 Intervals with a Small Imaginary Part

With a view towards later use we now consider the case of a manifold that is primarily real on which $g_{\mu\nu}$ has a small imaginary part, and where the imaginary part of infinitesimals dX^μ is much smaller than the real part. The quadratic form to first order in imaginary quantities is:

$$d\tau^2 = {}_rg_{\mu\nu}\,{}_rdX^\mu\,{}_rdX^\nu + i\,(2\,{}_rg_{\mu\nu}\,{}_rdX^\mu\,{}_idX^\nu + {}_ig_{\mu\nu}\,{}_rdX^\mu\,{}_rdX^\nu) \quad (5\text{-A.5.5.1})$$

As a result the values of $d\tau$ will cluster around the real or imaginary $d\tau$ axes (See Fig. 5-A.5.5.1.) depending on whether the real part of $d\tau^2$ is space-like or time-like resulting in a "fuzzy" version of the usual view of intervals as space-like or time-like.

5-A.6 Covariant Derivatives along a Line

We have previously defined the covariant derivative for covariant and contravariant vectors (eqs. 5-A.5.3.2 and 5-A.5.3.3) as well as more generally for tensors.

We will now examine the rate of change of a vector (and tensors) along a curve in complex space-time. A curve can be parameterized by the complex interval $\tau = \int d\tau$ integrated from the beginning point of the curve, or by the real valued interval $s = \int ds$ similarly measured from the beginning of the curve. The quantities τ and s along a space-time curve C are implicitly related through the expressions:

$$\tau = \int_C d\tau = \int_C ds\, d\tau/ds \quad (5\text{-A.5.1a})$$

$$s = \int_C ds \quad (5\text{-A.5.1b})$$

More particularly, since

$$d\tau/ds = e^{i\theta} \quad (5\text{-A.5.2})$$

if we use a polar representation of the complex infinitesimal $d\tau$:

$$d\tau = e^{i\theta}\, ds \quad (5\text{-A.5.3})$$

with θ being a phase angle that depends on the space-time point at which the interval is measured. Consequently,

$$\tau = \int_C ds\, e^{i\theta} \quad (5\text{-A.5.4})$$

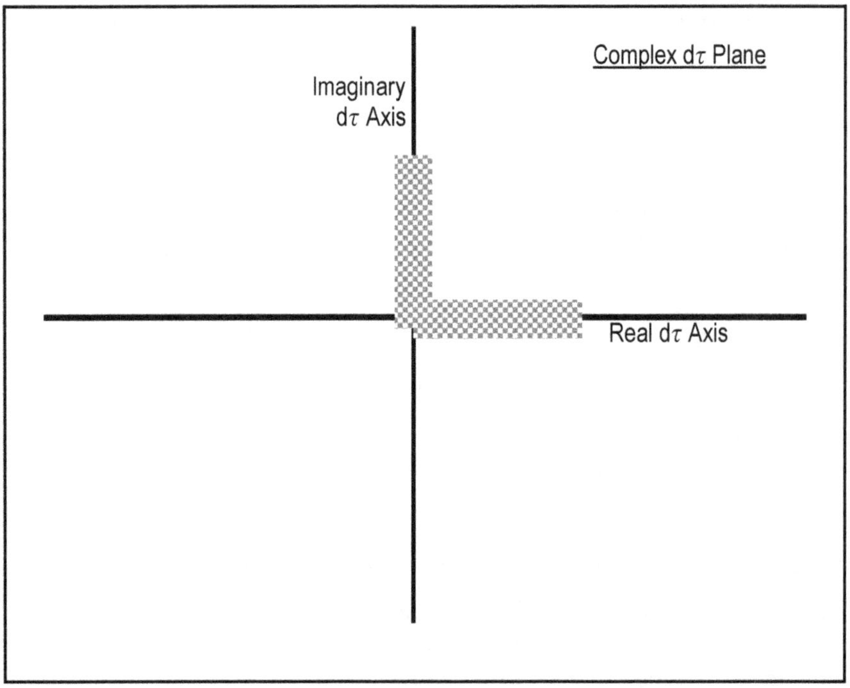

Figure 5-A.5.5.1. Values of dτ for the case of small imaginary parts.

If θ is constant along the curve then

$$\tau = e^{i\theta}\ s \tag{5-A.5.5}$$

More generally, it is easy to show

$$|\tau| \le s \tag{5-A.5.6}$$

The covariant derivative of a contravariant vector *along a curve* can be defined using the general expression for the covariant derivative of a contravariant vector:

$$Z^{\nu}_{;\mu} = \partial Z^{\nu}/\partial X^{\mu} + \Gamma^{\nu}_{\ \mu\lambda}\, Z^{\lambda} \tag{5-A.5.7}$$

We define the covariant derivative with respect to s along a curve C as

$$\delta Z^{\nu}/\delta_C s\ = Z^{\nu}_{;\mu}\, dX^{\mu}/ds \equiv Z^{\nu}_{;\mu}\, dX^{\mu}/d\tau\ \cdot d\tau/ds \tag{5-A.5.8}$$

using the chain rule, and the covariant derivative with respect to τ along a curve C as

$$\delta Z^{\nu}/\delta_{C}\tau = Z^{\nu}_{;\mu}\, dX^{\mu}/d\tau \qquad (5\text{-A.5.9})$$

Eqs. 5-A.5.8 and 5-A.5.9 can be used to determine the covariant derivative of particle variables, such as the momentum $p^{\alpha}(\tau)$ and spin $S^{\beta}(\tau)$, along a trajectory.

The covariant derivatives of a covariant vector along a curve or trajectory C are:

$$\delta Z_{\nu}/\delta_{C}s = Z_{\nu;\mu}\, dX^{\mu}/ds \equiv Z_{\nu;\mu}\, dX^{\mu}/d\tau \cdot d\tau/ds \qquad (5\text{-A.5.10})$$

and

$$\delta Z_{\nu}/\delta_{C}\tau = Z_{\nu;\mu}\, dX^{\mu}/d\tau \qquad (5\text{-A.5.11})$$

The generalization to covariant derivatives of higher rank tensors along a curve is straightforward.

5-A.7 Parallel Transport and Line Integrals

5-A.7.1 Parallel Transport

Parallel transport in complex space-time is similar to parallel transport in real space-time. The change in a covariant vector Z_{ν} when *parallel transported* to a nearby point is defined by

$$Z_{\nu;\mu} = 0 \qquad (5\text{-A.7.1.1})$$

Thus by eq. 5-A.5.3.3 the change in Z_{ν} when parallel transported by the infinitesimal distance dX^{μ} is:

$$dZ_{\nu} = \Gamma^{\lambda}{}_{\nu\mu}\, Z_{\lambda} dX^{\mu} \qquad (5\text{-A.7.1.2})$$

which can be re-expressed in terms of the complex interval $d\tau$ as

$$dZ_{\nu}/d\tau = \Gamma^{\lambda}{}_{\nu\mu}\, Z_{\lambda} dX^{\mu}/d\tau \qquad (5\text{-A.7.1.3})$$

and in terms of the physical interval ds as

$$dZ_{\nu}/ds = \Gamma^{\lambda}{}_{\nu\mu}\, Z_{\lambda} dX^{\mu}/d\tau \cdot d\tau/ds \qquad (5\text{-A.7.1.4})$$

5-A.7.2 Line Integrals in Complex Space-time

There are two types of line integrals in complex space-time: line integrals in the complex plane of a single coordinate, and line integrals in several coordinates in complex space-time. We will consider line integrals along a curve between two end points as well as line integrals around a closed curve.

5-A.7.3 Line Integrals in the Complex Plane of One Coordinate

Since each complex space-time coordinate can be viewed as defining a complex plane we can define a line integral in the complex plane of a single coordinate X^ν where $\nu = 0, 1, 2, 3$.

If we hold the other three complex coordinates constant then a line integral of a holomorphic function of the space-time coordinates X, where the curve is solely in the complex X^ν plane, becomes a conventional line integral of a holomorphic function of a single complex variable.

Consider the line integral of a function (or tensor) f(X) along a curve C in the X^ν complex plane between the points $X^\nu{}_0$ and $X^\nu{}_1$ as in Fig. 5-A.7.3.1:

$$I = \int_C f(X) dX^\nu \tag{5-A.7.3.1}$$

If f(X) is holomorphic in a simply connected domain D of the X^ν complex plane then Cauchy's Theorem implies

$$\oint_{C_D} f(X) dX^\nu = 0 \tag{5-A.7.3.2}$$

for any simple closed curve C_D in D. Thus the integral in eq. 5-A.7.3.1 between the points $X^\nu{}_0$ and $X^\nu{}_1$ is independent of the curve within the domain D and depends only on the endpoints. If there is a singularity such as a pole in D then the value of I in eq. 5-A.7.3.1 depends on the curve C, and an integral on a closed path C_D containing the singularity is not zero in general. Other results of the theory of complex variables also apply here without change.

5-A.7.4 Line Integrals in Several Coordinates in Complex Space-time

The question of line integrals in complex space-time that involve two or more of the complex dimensions is somewhat more complicated. Consider the line integral of a covariant vector $Z_\mu(X)$:

$$I_C = \int_C Z_\mu(X) dX^\mu \tag{5-A.7.4.1}$$

where the path C wanders through the complex planes of several complex variables from point X_1 to point X_2. If Z_μ is the gradient of a continuous regular function $Z_\mu = \partial_\mu \phi(X)$ then the integral is path independent and depends only on the end points in flat space-time where $g_{\mu\nu} = \eta_{\mu\nu}$

$$I_C = \phi(X_2) - \phi(X_1) \tag{5-A.7.4.2}$$

In a curved space-time the value of the integral I_C is path-dependent in general. However if Z_μ does not have any singularities in the complex planes of the space-time coordinates of the path traversed then I_C is independent of the path within the complex planes of the coordinates although it remains dependent on the overall path between the endpoints of the curve.

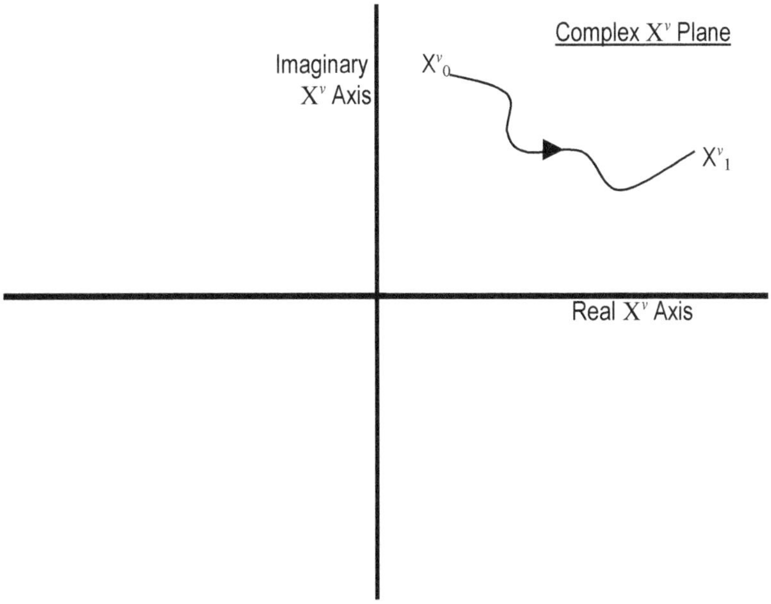

Figure 5-A.7.3.1 A curve solely in the complex X^ν plane extending from the coordinate value X^ν_0 to the coordinate value X^ν_1.

5-A.7.5 Line Integral in Several Complex Coordinates around a Closed Curve – Generalized Cauchy Theorem

In the theory of one complex variable, Cauchy's theorem states the line integral of a function around a closed curve C is zero if the function is holomorphic on C and within the region enclosed by C. We now ask whether Cauchy's theorem can be extended to complex space-time.

Consider a closed curve C in complex space-time. The curve C is assumed to follow a path that involves several complex dimensions (see Fig. 5-A.7.5.1).

Figure 5-A.7.5.1 A symbolic depiction of a closed curve in complex space-time.

A line integral around C can be viewed as a sum of line integrals around rectangular patches in a sufficiently fine grid (in the limit that the rectangles in the grid become infinitesimal in size.) See Fig. 5-A.7.5.2. The line integral contributions from the common interior lines cancel in pairs so that the sum of the line integral contributions of the infinitesimal rectangles is equal to the sum of the uncanceled contributions from the outer edges along C.

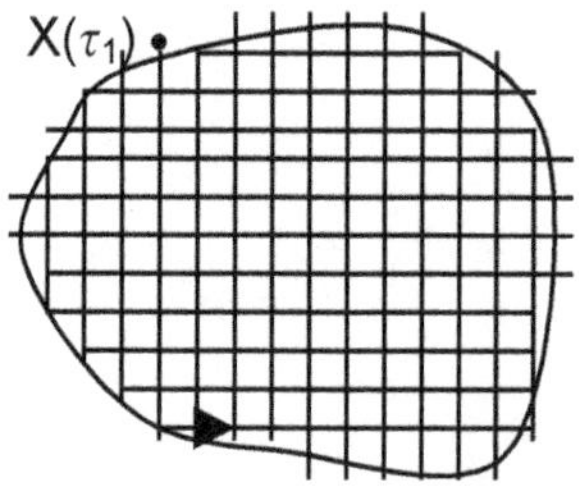

Figure 5-A.7.5.2 A closed curve in complex space-time. A grid of sufficiently fine mesh is superimposed on the area enclosed by the curve.

Thus the question of whether the integral of a holomorphic function around a closed space-time curve is zero or not can be reduced to a consideration of the line integral around an infinitesimal rectangle. To that end we will consider the line integral

$$I_c = \oint_c Z_\mu(X)dX^\mu \qquad (5\text{-A.7.5.1})$$

where c is a closed curve around an infinitesimal area. Let us assume that the point $X(\tau_1)$ lies on the curve. Then we can parallel transport Z_μ around c using eq. 5-A.7.1.3. We approximate the affine connection with the first few terms of its power series expansion:

$$\Gamma^\sigma{}_{\lambda\mu}(X(\tau)) = \Gamma^\sigma{}_{\lambda\mu}(X(\tau_1)) + (X^a(\tau) - X^a(\tau_1))\, \partial\Gamma^\sigma{}_{\lambda\mu}(X(\tau_1))/\partial X^a + \ldots \qquad (5\text{-A.7.5.2})$$

We expect the approximation to be good since c encloses an infinitesimal area. Substituting the expansion in eq. 5-A.7.1.3 we obtain an approximate expression for $Z_\mu(X(\tau))$:

$$Z_\mu(X(\tau)) \cong Z_\mu(X(\tau_1)) + (X^\lambda(\tau) - X^\lambda(\tau_1))\,\Gamma^\sigma{}_{\mu\lambda}(X(\tau_1))\, Z_\sigma(X(\tau_1)) \quad (5\text{-A.7.5.3})$$

up to first order in $X^\lambda(\tau) - X^\lambda(\tau_1)$. Combining eqs. 5-A.7.5.2 and 5-A.5.7.3, substituting in eq. 5-A.7.1.3, and forming an integral for we find $Z_\mu(X(\tau))$ to be:

$$Z_\mu(X(\tau)) \cong Z_\mu(X(\tau_1)) + \int_{\tau_1}^{\tau} d\tau \left\{\Gamma^\sigma{}_{\mu\rho}(X(\tau_1)) + (X^a(\tau) - X^a(\tau_1))\partial\Gamma^\sigma{}_{\mu\rho}(X(\tau_1))/\partial X^a\right\}\bullet$$

$$\bullet\left\{Z_\sigma(X(\tau_1)) + (X^\lambda(\tau) - X^\lambda(\tau_1))\,\Gamma^\beta{}_{\sigma\lambda}(X(\tau_1))\, Z_\beta(X(\tau_1))\right\} dX^\rho/d\tau \quad (5\text{-A.7.5.4})$$

to second order in $X^\lambda(\tau) - X^\lambda(\tau_1)$.

We now rewrite eq. 5-A.7.5.1 as an integral over a path parameterized by $X^\lambda(\tau)$:

$$I_c = \int_{\tau_1}^{\tau_2} d\tau\, Z_\mu(X)\, dX^\mu/d\tau \quad (5\text{-A.7.5.5})$$

where the parameter value τ_2 will close the curve: $X^\lambda(\tau_2) = X^\lambda(\tau_1)$ for $\lambda = 0, 1, 2, 3$. Substituting eq. 5-A.7.5.4 in eq. 5-A.7.5.5 gives

$$I_c \cong \left\{\Gamma^\sigma{}_{\mu\rho}(X(\tau_1))\Gamma^\beta{}_{\sigma a}(X(\tau_1)) + \partial\Gamma^\beta{}_{\mu\rho}(X(\tau_1))/\partial X^a\right\} Z_\beta(X(\tau_1))\, J^{\mu a\rho}{}_c \quad (5\text{-A.7.5.6})$$

to second order in $X^\lambda(\tau) - X^\lambda(\tau_1)$ after discarding some terms that become zero for the closed curve, The factor $J^{\mu a\rho}{}_c$ is given by

$$J^{\mu a\rho}{}_c = \int_{\tau_1}^{\tau_2} d\tau'\, dX^\mu/d\tau' \int_{\tau_1}^{\tau'} d\tau (X^a(\tau) - X^a(\tau_1))\, dX^\rho/d\tau \quad (5\text{-A.7.5.7})$$

which becomes

$$J^{\mu a\rho}{}_c = \int_{\tau_1}^{\tau_2} d\tau\, dX^\rho/d\tau [X^a(\tau) - X^a(\tau_1)][X^\mu(\tau_2) - X^\mu(\tau)] \quad (5\text{-A.7.5.8})$$

after switching the order of integration.

After some further simple algebra we find $J^{\mu a\rho}{}_c$ has an anti-symmetric term *and a symmetric term* in the indices a and ρ:

$$J^{\mu a\rho}{}_c = {}_A J^{\mu a\rho}{}_c + {}_S J^{\mu a\rho}{}_c \quad (5\text{-A.7.5.9})$$

where

$$ {}_{A}J^{\mu\alpha\rho}{}_{c} = X^{\mu}(\tau_2)\int_{\tau_1}^{\tau_2} d\tau\, dX^{\rho}/d\tau\, X^{\alpha}(\tau) + \tfrac{1}{2}\int_{\tau_1}^{\tau_2} d\tau\, X^{\mu}[dX^{\rho}/d\tau\, X^{\alpha}(\tau) - dX^{\alpha}/d\tau\, X^{\rho}(\tau)] + $$

$$ + \tfrac{1}{2}\int_{\tau_1}^{\tau_2} d\tau\, X^{\mu}(\tau)[X^{\alpha}(\tau_1)\, dX^{\rho}/d\tau - X^{\rho}(\tau_1)\, dX^{\alpha}/d\tau] \quad (5\text{-A.7.5.10}) $$

and

$$ {}_{S}J^{\mu\alpha\rho}{}_{c} = -\tfrac{1}{2}\int_{\tau_1}^{\tau_2} d\tau\, X^{\rho}\, X^{\alpha}\, dX^{\mu}/d\tau + \tfrac{1}{2}\int_{\tau_1}^{\tau_2} d\tau\, X^{\mu}(\tau)[X^{\alpha}(\tau_1)dX^{\rho}/d\tau + X^{\rho}(\tau_1)dX^{\alpha}/d\tau] \quad (5\text{-A.7.5.11}) $$

Eq. 5-A.7.5.6 can now be written

$$ I_c \cong \tfrac{1}{2}[R^{\beta}{}_{\mu\rho\alpha}\, {}_{A}J^{\mu\alpha\rho}{}_{c} + S^{\beta}{}_{\mu\rho\alpha}\, {}_{S}J^{\mu\alpha\rho}{}_{c}]\, Z_{\beta}(X(\tau_1)) \quad (5\text{-A.7.5.12}) $$

where $R^{\beta}{}_{\mu\rho\alpha}$ is the Riemann-Christoffel curvature tensor (eq. 5-A.5.4.1) and $S^{\beta}{}_{\mu\rho\alpha}$ is a non-tensor which we will call the *symmetric curvature*:

$$ S^{\beta}{}_{\mu\rho\alpha} = \partial\Gamma^{\beta}{}_{\mu\rho}/\partial X^{\alpha} + \partial\Gamma^{\beta}{}_{\mu\alpha}/\partial X^{\rho} + \Gamma^{\sigma}{}_{\mu\rho}\Gamma^{\beta}{}_{\sigma\alpha} + \Gamma^{\sigma}{}_{\mu\alpha}\Gamma^{\beta}{}_{\sigma\rho} \quad (5\text{-A.7.5.13}) $$

because it is symmetric in α and ρ. It does *not* have the index symmetries of $R^{\beta}{}_{\mu\rho\alpha}$ (eqs. 5-A.5.4.6 – 5-A.5.4.8).

From eq. 5-A.7.5.12 we see that the line integral around a closed curve in several coordinates in complex space-time is not zero unless the curvature tensor and the symmetric curvature are both zero. In a flat space-time both are zero if the metric tensor is constant such as in the case of cartesian coordinates or light cone coordinates. If the metric tensor $g_{\mu\rho}$ is not constant such as in the case of a spherical spatial coordinate system, then $S^{\beta}{}_{\mu\rho\alpha}$ is non-zero and line integrals around closed curves are non-zero in general. In a curved space-time both $R^{\beta}{}_{\mu\rho\alpha}$ and $S^{\beta}{}_{\mu\rho\alpha}$ are non-zero in general.

5-A.7.5.1 Parallel Transport of a Vector around a Closed Curve in Several Variables

The vector $Z_\mu(X(\tau))$ itself, given in eq. 5-A.7.5.4, when parallel transported around an infinitesimal closed curve at $X(\tau_1)$ will be unchanged if and only if the Riemann-Christoffel tensor $R^\beta{}_{\mu\rho\alpha}$ is zero at $X(\tau_1)$. Starting from eq. 5-A.7.5.4 it is easy to show that the net change is

$$\Delta Z_\beta = Z_\beta(X(\tau_2)) - Z_\beta(X(\tau_1)) \cong \tfrac{1}{2} R^\alpha{}_{\beta\rho\sigma}(\tau_1) Z_\alpha(X(\tau_1)) \oint d\tau\, X^\sigma dX^\rho/d\tau \tag{5-A.7.5.1.1}$$

Thus the change in the vector does not depend on the symmetric curvature $S^\beta{}_{\mu\rho\alpha}$.

5-A.7.5.2 Generalization of Cauchy's Theorem to Curved, Complex Space-time

In section 5-A.7.3 we considered line integrals in the complex plane of one coordinate and concluded that a line integral of a function around a closed curve in the complex plane of a single coordinate was zero if the curvature is zero and the function is holomorphic in the coordinate on, and inside, the curve (eq. 5-A.7.5-A.2). Eq. 5-A.7.5.12 enables us to make this statement more precise.

Suppose we consider the case of a vector function $Z_\mu(X(\tau))$ integrated around an infinitesimal closed curve c solely in the complex plane of the coordinate $X^{\mu 0}$ as developed from eq. 5-A.7.5.5. We assume the vector $Z_\mu(X)$ is holomorphic on c and on the surface enclosed by c. Then eq. 5-A.7.5.5 becomes

$$I_c = \int_{\tau_1}^{\tau_2} d\tau\, Z_{\mu_0}(X)\, dX^{\mu_0}/d\tau \tag{5-A.7.5.2.1}$$

with no implied sum over μ_0. Following the same line of development as section 5-A.7.5 we arrive at a special case of eq. 5-A.7.5.12 using the analyticity of Z_μ, and of the affine connection, to obtain the equivalent of eq. 5-A.7.5.4 and then to obtain

$$I_c \cong \tfrac{1}{2}[R^\beta{}_{\mu_0\rho\alpha}\, {}_A J^{\mu_0\alpha\rho}{}_c + S^\beta{}_{\mu_0\rho\alpha}\, {}_S J^{\mu_0\alpha\rho}{}_c]\, Z_\beta(X(\tau_1)) \tag{5-A.7.5.2.2}$$

The line integral of an holomorphic vector around a closed infinitesimal curve in the complex plane of a single coordinate depends on the Riemann-Christoffel curvature tensor and the symmetric curvature. It is zero in a flat complex space-time. Thus we obtain a generalization of Cauchy's Theorem in curved complex space-time:

5-A.7.5.2.1 ONE COMPLEX VARIABLE CAUCHY THEOREM IN CURVED COMPLEX SPACE-TIME

$$\int_c dX^{\mu_0} Z_{\mu_0}(X) = \tfrac{1}{2}[R^{\beta}{}_{\mu_0 \rho a}\, {}_A J^{\mu_0 a\rho}{}_c + S^{\beta}{}_{\mu_0 \rho a}\, {}_S J^{\mu_0 a\rho}{}_c]\, Z_\beta(X(\tau_1)) \tag{5-A.7.5.2.1.1}$$

with no sum over μ_0.

5-A.7.5.2.2 SEVERAL COMPLEX VARIABLES CAUCHY THEOREM IN CURVED COMPLEX SPACE-TIME

We also obtain the several complex variables generalization:

$$\int_c dX^{\mu} Z_{\mu}(X) = \tfrac{1}{2}[R^{\beta}{}_{\mu \rho a}\, {}_A J^{\mu a\rho}{}_c + S^{\beta}{}_{\mu \rho a}\, {}_S J^{\mu a\rho}{}_c]\, Z_\beta(X(\tau_1)) \tag{5-A.7.5.2.2.1}$$

with an implied sum over μ.

5-A.7.5.3 Parallel Transport of a Vector around a Closed Curve in the Complex Plane of a Coordinate

In section 5-A.7.5.3 we examined the line integral of a vector around a closed curve in the complex plane of one coordinate X^{μ_0}. In this section we note that the parallel transport of a vector around a closed curve in the complex plane of a coordinate may be obtained directly from eq. 5-A.7.5.4. After some algebra we find

$$\Delta Z_\beta = Z_\beta(X(\tau_2)) - Z_\beta(X(\tau_1)) \cong [\partial \Gamma^{a}{}_{\beta\mu_0}/\partial X^{\mu_0} + \Gamma^{\nu}{}_{\beta\mu_0}\Gamma^{a}{}_{\nu\mu_0}] Z_a(X(\tau_1)) \oint X^{\mu_0} dX^{\mu_0}$$
$$= 0 \tag{5-A.7.5.3.1}$$

(no sum over the μ_0 index) due to the integral around the closed path being zero. Eq. 5-A.7.5.3.1 is based on expansions (eqs. 5-A.7.5.2 and 3.7.5.3) that require analyticity in Z_μ and the affine connection.

5-A.7.5.4 Path Independence of Parallel Transported Vectors in Complex Space-time

In section 5-A.7.5.1 we examined the parallel transport of a vector around a closed infinitesimal curve and found that the change in the vector was zero if the Riemann-Christoffel curvature tensor was zero in the neighborhood of the closed curve (eq. 5-A.7.5.1.1). By combining closed infinitesimal curves in a grid pattern one can construct a finite curve for which the conclusion also applies; namely, *that the parallel transport of a vector around a closed finite curve leaves the vector unchanged if the Riemann-Christoffel curvature tensor is zero in a domain D containing the curve.*

Therefore the parallel transport of a vector between two points in D, X_1 and X_2, is independent of the choice of curve (within D) between X_1 and X_2. The components of the vector at X_2 are the same independent of the curve (within D) along which the vector is transported *if the Riemann-Christoffel curvature tensor is zero in D.*

5-A.7.5.5 Path Independence of Line Integrals of Vectors in Complex Space-time

In section 5-A.7.5 we examined the line integral of a vector around a closed infinitesimal curve and found that the value of the line integral was zero if the Riemann-Christoffel curvature tensor and the symmetric curvature were both zero in the neighborhood of the closed curve (eq. 5-A.7.5.12).

Again we can combine closed infinitesimal curves in a grid pattern to construct a line integral around a finite curve for which the conclusion also applies; namely, *that the line integral of a vector around a closed finite curve is zero if the Riemann-Christoffel curvature tensor and the symmetric curvature are both zero in a domain D containing the curve (eq. 5-A.7.5.2.2.1).*

Therefore the value of the line integral of a vector between two points in D, X_1 and X_2, is independent of the choice of curve (within D) between X_1 and X_2. *if the Riemann-Christoffel curvature tensor and the symmetric curvature are both zero in D.*

Since the symmetric curvature is not a tensor, a change of coordinate system can transform it from a zero to a non-zero value or vice versa. If the Riemann-Christoffel curvature tensor is zero in one coordinate system it remains zero when a transformation to another coordinate system is made.

Appendix 5-B. Integration in Complex Space-time & an Anti-Holomorphic Extension of Space-time

This appendix may be skipped by the reader since it is not important for the remaining chapters.

5-B.1 Eight Dimensional Real Representation of Complex Space-time

Up to this point the formulation of the geometry of complex space-time has solely used complex holomorphic coordinates z^μ where $\mu = 0, 1, 2, 3$. In particular we have used holomorphic functions and variables in the differential equations and line integrals that describe our complex curved space-time.

At this point we wish to consider volume and surface integrals in complex space-time. This development will lead us to introduce anti-holomorphic coordinates and quantities. Physically, we begin with definition of a volume integral in terms of the real and imaginary parts of each coordinate:

$$I = \int dx^0 dy^0 dx^1 dy^1 dx^2 dy^2 dx^3 dy^3 \, f(x^0,y^0,x^1,y^1,x^2,y^2,x^3,y^3) \tag{5-B.1.1}$$

where $f(x^0,y^0,x^1,y^1,x^2,y^2,x^3,y^3)$ is an arbitrary function, and where each complex coordinate is expressed in terms of its real and imaginary parts:

$$X^\mu = x^\mu + iy^\mu \tag{5-B.1.2}$$

We now introduce the complex conjugate coordinates (variables), which we will call the *anti-holomorphic coordinates* for short:

$$X^{\mu *} = x^\mu - iy^\mu \tag{5-B.1.3}$$

In addition, we define differential operators corresponding to these variables:

$$\partial/\partial X^\mu = \partial/\partial x^\mu - i\partial/\partial y^\mu \tag{5-B.1.4}$$

$$\partial/\partial X^{\mu *} = \partial/\partial x^\mu + i\partial/\partial y^\mu \tag{5-B.1.5}$$

All the formal rules of differentiation of real variables also hold for these operators. In addition it is easy to verify that they are independent in the sense that:

$$\partial X^{\nu}/\partial X^{\mu *} = \partial X^{\nu *}/\partial X^{\mu} = 0 \qquad (5\text{-B.1.6})$$

for all μ and ν.

A major consequence of the formal identity of the rules for real and complex differentiation is:

Rule 5-B.1.1. If a differentiable function g(X) is defined in a domain D that includes a section of the real subspace D_R ($y^{\mu} = 0$ for all μ), and if g(X) is equal to a real function g(x) at all points $z^{\mu} = x^{\mu}$ for μ=0, 1, 2,3 of D_R, then the partial derivatives of g(X) will equal the corresponding derivatives of g(x) to all orders in the domain D_R. This rule applies to both scalar and tensor functions.

Thus, in particular, we can complexify Riemannian space-time and general relativity by simply substituting complex coordinates in all equations that are based solely on the metric $g_{\mu\nu}$, and its derivatives, and algebraic combinations thereof. Integral relations must be handled on a case-by-case basis as we will see later in this apendix.

Eq. 5-B.1.1 can be rewritten in terms of X^{μ} and $X^{\mu *}$ as:

$$I = \int dX^0 dX^{0*} dX^1 dX^{1*} dX^2 dX^{2*} dX^3 dX^{3*}\ \partial(x,y)/\partial(X,X^*)$$
$$f(X^0,X^{0*},X^1,X^{1*},X^2,X^{2*},X^3,X^{3*}) \qquad (5\text{-B.1.7})$$

where

$$\partial(x,y)/\partial(X,X^*) \equiv \partial(x^0,y^0,x^1,y^1,x^2,y^2,x^3,y^3)/\partial(X^0,X^{0*},X^1,X^{1*},X^2,X^{2*},X^3,X^{3*})$$
$$(5\text{-B.1.8})$$

is the Jacobian of the transformation from the eight real variables of eq. 5-B.1.1 to the four holomorphic and four anti-holomorphic variables in eq. 5-B.1.7.

5-B.2 Relation of 8-Dimensional Real Space-time Jacobians to Complex Space-time Jacobians

Since our development of complex space-time features using the complex coordinates X^{μ} is concise and convenient we will also develop volume and surface integration using them. In order to establish the form of complex integration we will need to relate the Jacobian for a change of coordinate system in the x^{μ} and y^{μ} variables to the Jacobian for a change of coordinate system in the X^{μ} and $X^{*\mu}$ variables:

$$\partial(x,y)/\partial(x',y') = \partial(x^0,y^0,x^1,y^1,x^2,y^2,x^3,y^3)/\partial(x'^0,y'^0,x'^1,y'^1,x'^2,y'^2,x'^3,y'^3)$$
$$(5\text{-B.2.1})$$

and

$$\partial(X,X^*)/\partial(X',X'^*) = \partial(X^0,X^{0*},X^1,X^{1*},X^2,X^{2*},X^3,X^{3*})/\partial(X'^0,X'^{0*},X'^1,X'^{1*},X'^2,X'^{2*},X'^3,X'^{3*}) \qquad (5\text{-B.2.2})$$

We will show

$$\partial(x,y)/\partial(x',y') = \partial(X,X^*)/\partial(X',X'^*) \qquad (5\text{-B.2.3})$$

if

$$X'^\mu = x'^\mu + iy'^\mu \qquad (5\text{-B.2.4})$$

and

$$X'^{\mu*} = x'^\mu - iy'^\mu \qquad (5\text{-B.2.5})$$

Proof
First we note that

$$\partial(X,X^*)/\partial(x,y) = (-2i)^4 \qquad (5\text{-B.2.6})$$

using

$$\partial X^\mu/\partial x^\nu = \delta^\mu{}_\nu \qquad \partial X^\mu/\partial y^\nu = i\delta^\mu{}_\nu \qquad (5\text{-B.2.7})$$

and

$$\partial X^{\mu*}/\partial x^\nu = \delta^\mu{}_\nu \qquad \partial X^{\mu*}/\partial y^\nu = -i\delta^\mu{}_\nu \qquad (5\text{-B.2.8})$$

where $\delta^\mu{}_\nu = 1$ if $\mu = \nu$ and zero otherwise. Similarly

$$\partial(X',X'^*)/\partial(x',y') = (-2i)^4 \qquad (5\text{-B.2.9})$$

Comparing eqs. 5-B.2.6 and 5-B.2.9 we see

$$\partial(X,X^*)/\partial(x,y) = \partial(X',X'^*)/\partial(x',y') \qquad (5\text{-B.2.10})$$

Multiplying both sides of eq. 5-B.2.10 by $\partial(x,y)/\partial(X',X'^*)$ and using the multiplication rule for determinants det(AB) = det(A)det(B) we find the result eq. 5-B.2.3.■

5-B.3 Coordinate Transformations in 4 Complex Dimensions

In section 5-A.4 we defined a complex coordinate transformations with a set of continuous, holomorphic functions f^μ:

$$X^\mu = f^\mu(X'^0, X'^1, X'^2, X'^3) \qquad (5\text{-A.4.1})$$

for $\mu = 0, 1, 2, 3$ where the coordinates X'^ν are also holomorphic. We note the partial derivatives of X^μ satisfy:

$$\partial X^{\mu}/\partial X'^{\nu} \equiv \partial f^{\mu}(X'^{0}, X'^{1}, X'^{2}, X'^{3})/\partial X'^{\nu} \neq 0 \qquad (5\text{-B.3.2})$$

in general; while

$$\partial X^{\mu}/\partial X'^{\nu *} \equiv \partial f^{\mu}(X'^{0}, X'^{1}, X'^{2}, X'^{3})/\partial X'^{\nu *} = 0 \qquad (5\text{-B.3.3})$$

due to the holomorphy of f^{μ} in each coordinate.

5-B.4 8-Dimensional Complex Space-time

We now extend our four-dimensional complex space-time to an eight-dimensional complex space-time in order to more conveniently express volume and surface integrals in complex space-time. The additional four dimensions are defined using the complex conjugate of eq. 5-A.4.1. Thus these extra dimensions, as we use them, do not introduce new degrees of freedom. Therefore, strictly speaking, we are dealing with a subspace of a full, eight-dimensional complex space-time. We define the additional coordinates via:

$$X^{*\mu} = f^{\mu - 4 *}(X'^{0*}, X'^{1*}, X'^{2*}, X'^{3*}) \qquad (5\text{-A.4.1})$$

for $\mu = 4, 5, 6, 7$. (An extension of this approach, which is not pursued in this book, would be to define a set of independent functions f^{a} for $a = 4, 5, 6, 7$.) Thus the coordinates X^{ν} and X'^{ν} are the complex conjugates of $X^{\nu-4}$ and $X'^{\nu-4}$ respectively for $\nu = 4, 5, 6, 7$:

$$X^{4} = X^{0*} \qquad (5\text{-B.4.1a})$$
$$X^{5} = X^{1*} \qquad (5\text{-B.4.1b})$$
$$X^{6} = X^{2*} \qquad (5\text{-B.4.1c})$$
$$X^{7} = X^{3*} \qquad (5\text{-B.4.1d})$$

and

$$X'^{4} = X'^{0*} \qquad (5\text{-B.4.2a})$$
$$X'^{5} = X'^{1*} \qquad (5\text{-B.4.2b})$$
$$X'^{6} = X'^{2*} \qquad (5\text{-B.4.2c})$$
$$X'^{7} = X'^{3*} \qquad (5\text{-B.4.2d})$$

In the interests of simplicity we will simply use the complex conjugates of coordinates $X^{\nu *}$ and $X'^{\nu *}$ for $\nu = 0, 1, 2, 3$ in the following discussions.

5-B.5 8-Dimensional Complex Coordinate Transformations

We can use a matrix notation to establish the form of complex coordinate transformations. We start by defining 4-dimensional complex coordinate transformations in matrix form. Then we define a set of 8-dimensional complex coordinate transformations. Our set of transformations is a subset of the complete set of 8-dimensional complex coordinate transformations.

5-B.5.1 Matrix Representation of 4 Complex Dimensions Coordinate Transformations

Let us begin by considering a local transformation in 4-dimensional complex space-time that transforms the flat space metric tensor $\eta_{\alpha\beta}$ where X^{μ} represents the flat space coordinates to a new form $g_{\mu\nu}$ in the coordinate system X'^{μ}

$$g_{\mu\nu}(X') = \eta_{\alpha\beta}\,\partial X^{\alpha}/\partial X'^{\nu}\;\partial X^{\beta}/\partial X'^{\nu} \tag{5-B.5.1.1}$$

This can be expressed in matrix form as

$$\left(g_{\mu\nu}(X')\right) = \Lambda^{T}\left(\partial X^{\alpha}/\partial X'^{\nu}\right)\left(\eta_{\alpha\beta}\right)\Lambda\left(\partial X^{\beta}/\partial X'^{\nu}\right) \tag{5-B.5.1.2}$$

where Λ is a complex (generally) matrix formed of partial derivatives, and Λ^{T} is its transpose. We will write eq. 5-B.5.1.2 in the abbreviated form:

$$(g) = (\Lambda^{T})(\eta)(\Lambda) \tag{5-B.5.1.3}$$

5-B.5.2 Matrix Representation of 8 Complex Dimensions Coordinate Transformations

We now use the matrices defined in the preceding section to define an 8-dimensional matrix formulation. We define an 8-dimensional Minkowski metric matrix with

$$(\eta_8) = \begin{bmatrix} (\eta) & (0) \\ (0) & (\eta) \end{bmatrix} \tag{5-B.5.2.1}$$

with the usual 4 by 4 Minkowski metric matrices along the diagonal. We then define the 8-dimensional transformation matrix in terms of the 4-dimensional transformation matrix Λ.

$$(\Lambda_8) = \begin{bmatrix} (\Lambda) & (0) \\ (0) & (\Lambda^*) \end{bmatrix} \tag{5-B.5.2.2}$$

and the 8-dimensional metric $g_{8\mu\nu}$ with

$$(g_8) = \begin{bmatrix} (g) & (0) \\ (0) & (g^*) \end{bmatrix} \tag{5-B.5.2.3}$$

We then can write the 8-dimensional version of eq. 5-B.5.3:

$$(g_8) = (\Lambda_8^{\,T})(\eta_8)\,(\Lambda_8) \tag{5-B.5.2.4}$$

which implies eq. 5-B.5.1.3 and its complex conjugate:

$$(g^*) = (\Lambda^{T*})(\eta)\,(\Lambda^*) \tag{5-B.5.2.5}$$

since $\eta = \eta^*$.

5-B.6 The Complex Space-time Determinant g

In this section we evaluate the complex space-time determinant g. First we note:

$$I = \int dX^0 dX^{0*} dX^1 dX^{1*} dX^2 dX^{2*} dX^3 dX^{3*} f(X^0,X^{0*},X^1,X^{1*},X^2,X^{2*},X^3,X^{3*})$$

$$\equiv \int d^4X d^4X^*\; f(X,X^*) = \int d^4X' d^4X'^* \partial(X,X^*)/\partial(X',X'^*)\; f(X,X^*)$$

where

$$\partial(X,X^*)/\partial(X',X'^*) = \det \begin{bmatrix} (\partial X^\mu/\partial X'^\nu) & (0) \\ (0) & (\partial X^{\mu*}/\partial X'^{\nu*}) \end{bmatrix}$$

$$= \det(\Lambda_8) = \det(\Lambda)[\det(\Lambda)]^* \qquad (5\text{-B.5.1})$$

The matrix equations in section 5-B.5 enable us to relate the determinant of the metric tensor to the local transformation relating $g_{\mu\nu}$ to $\eta_{\alpha\beta}$. Eq. 5-B.5.1.3 implies

$$\det(g) = \det(\Lambda^T)\det(\eta)\det(\Lambda) \qquad (5\text{-B.5.1})$$

and thus gives

$$g = \det(g) = -[\det(\Lambda)]^2 \qquad (5\text{-B.5.2})$$

using $\det(\Lambda^T) = \det(\Lambda)$ and $\det(\eta) = -1$. Thus

$$g = -[\det(\Lambda)]^2 = -[\partial(X^0, X^1, X^2, X^3)/\partial(X'^0, X'^1, X'^2, X'^3)]^2 \qquad (5\text{-B.5.3})$$

is minus the square of a Jacobian just as in the case of real space-time. The complex conjugate g^* is determined from the complex conjugate of eq. 5-B.5.3

$$g^* = -[\det(\Lambda)]^{2*} = -[\partial(X^{0*}, X^{1*}, X^{2*}, X^{3*})/\partial(X'^{0*}, X'^{1*}, X'^{2*}, X'^{3*})]^2 \qquad (5\text{-B.5.4})$$

If we take the determinant of eq. 5-B.5.2.4 we find

$$g_8 \equiv \det(g_8) = [\det(\Lambda_8)]^2 = [\det(\Lambda)]^2[\det(\Lambda)]^{2*} = gg^* \qquad (5\text{-B.5.5})$$

using eq. 5-B.5.1.

5-B.7 Volume Integrals using Complex Coordinates

Eq. 5-B.1.1 defines a volume integral in a real, 8-dimensional local Lorentz frame. Eq. 5-B.1.7 re-expresses that integral in terms of complex coordinates:

$$I = \int dX^0 dX^{0*} dX^1 dX^{1*} dX^2 dX^{2*} dX^3 dX^{3*}\, \partial(x,y)/\partial(X,X^*)\; f(X^0,X^{0*},X^1,X^{1*},X^2,X^{2*},X^3,X^{3*}) \qquad (5\text{-B.1.7})$$

which becomes

$$I = \int dX^0 dX^{0*} dX^1 dX^{1*} dX^2 dX^{2*} dX^3 dX^{3*} f(X^0,X^{0*},X^1,X^{1*},X^2,X^{2*},X^3,X^{3*})/16 \qquad (5\text{-B.7.1})$$

in the local Lorentz frame using the inverse of eq. 5-B.2.5. If we transform to a different complex coordinate system then we find that the transformation leads to

$$dX^0dX^{0*}dX^1dX^{1*}dX^2dX^{2*}dX^3dX^{3*} = \partial(X,X^*)/\partial(X',X'^*)dX'^0dX'^{0*}\cdot \cdot dX'^1dX'^{1*}dX'^2dX'^{2*}dX'^3dX'^{3*} \quad (5\text{-B.7.2})$$

with $\partial(X,X^*)/\partial(X',X'^*)$ defined by eq. 5-B.2.2. Using eqs. 5-B.5.1 and 5-B.5.6 in conjunction with eqs. 5-B.7.1 and 5-B.7.2 gives

$$I = (\tfrac{1}{2}\, i)^{-4}\int dX'^0dX'^{0*}dX'^1dX'^{1*}dX'^2dX'^{2*}dX'^3dX'^{3*}(gg^*)^{1/2}f(X') \quad (5\text{-B.7.3})$$

which we abbreviate to

$$I = 2^{-4}\int d^4X'd^4X'^*(gg^*)^{1/2}f(X') \quad (5\text{-B.7.4})$$

Thus the *proper volume element* in our complex space-time formulation is

$$2^{-4}d^4Xd^4X^*\ (gg^*)^{1/2} \quad (5\text{-B.7.5})$$

5-B.8 Gauss' Theorem for Complex Space-time

Gauss' Theorem plays an important role in the analysis of features of general relativity. In this section we will derive the complex space-time version of Gauss' Theorem. It is significantly different from the real space-time equivalent.

Consider the integral of the covariant divergence of a holomorphic vector function F of the *holomorphic coordinates* X^μ: $F^\kappa(X)$ over a volume V. (We call the complex conjugate of the X^μ coordinates *anti-holomorphic coordinates* and denote them as $X^{\mu *}$.)

$$I = 2^{-4}\int_V d^4Xd^4X^*\ (gg^*)^{1/2}\ F^\rho{}_{;\rho} \quad (5\text{-B.8.1})$$

Now

$$F^\rho{}_{;\rho} = g^{-1/2}\ \partial(g^{1/2}F^\rho)/\partial X^\rho \quad (5\text{-B.8.2})$$

by eq. 5-A.5.3.8. Therefore eq. 5-B.8.1 becomes

$$I = 2^{-4}\int_V d^4Xd^4X^*\ g^{*1/2}\ \partial(g^{1/2}F^\rho)/\partial X^\rho \quad (5\text{-B.8.3})$$

Since the integral over the X variables is an exact divergence of F^ρ with respect to X^ρ we can use a variation of the complex form of Green's Theorem to eliminate one of the integrations over X thus producing a "surface" integral. The one complex variable Green's Theorem that we will use is:

Alternate One Complex Variable Version of Green's Theorem
If F(z, z*) is continuous and has continuous partial derivatives in some region R and on its boundary curve C then

$$\int_R dz dz^* \, \partial F/\partial z = \oint_C dz^* \, F(z, z^*) \tag{5-B.8.4}$$

Applying this theorem to each of the four terms in the integrand of eq. 5-B.8.3 we obtain:

$$I = 2^{-4} \int d\Sigma_\rho \, (gg^*)^{1/2} \, F^\rho \tag{5-B.8.5}$$

where

$$\int_V d\Sigma_\rho \equiv \int_V \prod_{\mu \neq \rho} dX^\mu dX^{\mu *} \oint_{C_\rho} dX_\rho{}^* \tag{5-B.8.6}$$

with C_ρ the curve bounding the area of the X_ρ integration and the other integrals spanning the volume of integration in those variables. Thus we have the

5-B.8.1 Complex Space-time Gauss' Theorem

$$\int_V d^4X d^4X^* (gg^*)^{1/2} F^\rho{}_{;\rho} = \int_V d\Sigma_\rho \, (gg^*)^{1/2} F^\rho \tag{5-B.8.1.1}$$

In addition the right side of eq. 5-B.8.1.1 displays the form of a *proper surface integral*. The next section illustrates the application of the Complex Space-time Gauss' Theorem. The generalization to higher dimensional complex spaces is direct.

5-B.9 Examples of Gauss' Theorem in Complex Space-time

In this section we will consider some examples of the complex space-time Gauss' Theorem.

5-B.9.1 Four-dimensional, Flat, Complex Space-time Gauss' Theorem Example: a Hypercube

We will consider the volume integration of the divergence of a vector, $V^\rho(X)$, for the case of a 4-dimensional hypercube with sides of length a in flat space-time ($g = g^* = -1$). Let

$$V^{\rho}(X) = X^{\rho} \tag{5-B.9.1.1}$$

The volume integral by direct integration of the real and imaginary parts is

$$I = \int dx^0dy^0dx^1dy^1dx^2dy^2dx^3dy^3\ (gg^*)^{1/2}V^{\rho}{}_{;\rho} = 4a^8 \tag{5-B.9.1.2}$$

The integration of the left side of the complex space-time Gauss' Theorem (eq. 5-B.8.1.1) multiplied by 2^{-4} is

$$I = 2^{-4}\int d^4Xd^4X^*\ (gg^*)^{1/2}\ V^{\rho}{}_{;\rho} = 2^{-4}\int d^4Xd^4X^*4 \tag{5-B.9.1.3}$$

$$= 2^{-2}\Big[\int dzdz^*\Big]^4 = 2^{-2}(-2ia^2)^4 = 4a^8 \tag{5-B.9.1.4}$$

The integration of the right side of the complex space-time Gauss' Theorem (eq. 5-B.8.1.1) multiplied by 2^{-4} is

$$J = 2^{-4}\int d\Sigma_{\rho}\ (gg^*)^{1/2}\ V^{\rho} = 2^{-4}\int \prod_{\mu\neq\rho} dX^{\mu}dX^{\mu *}\oint_{C_{\rho}} dX_{\rho}{}^*\ X^{\rho} \tag{5-B.9.1.5}$$

$$J = 2^{-4}\int\sum_{\rho}\prod_{\mu\neq\rho} dX^{\mu}dX^{\mu *}(-2ia^2) \tag{5-B.9.1.6}$$

using the complex variable result:

$$\oint dz^*\ z = -2iA \tag{5-B.9.1.7}$$

where A is the area enclosed by the curve for each of the 4 terms in the sum in eq. 5-B.9.1.5. For each term $A = a^2$. Performing the remaining integrals in eq. 5-B.9.1.6 yields:

$$J = 2^{-4}(-2ia^2)4(-2ia^2)^3 = 4a^8 = I \tag{5-B.9.1.8}$$

confirming the theorem.

5-B.9.2 Two-dimensional, Flat, Complex Space-time Gauss' Theorem Example: a Complex Sphere

We will consider the volume integration of the divergence of a vector, $V^{\rho}(X)$, for the case of a sphere in a flat, 2-dimensional complex Euclidean space with radius r. ($g = g^* = -1$). Let

$$V^{\rho}(X) = X^{\rho} \tag{5-B.9.2.1}$$

again. The volume integral by direct integration of the real and imaginary parts is

$$I = \int dx^1 dy^1 dx^2 dy^2\ (gg^*)^{1/2} V^{\rho}{}_{;\rho} = \pi^2 r^4 \tag{5-B.9.2.2}$$

The integration of the left side of the complex space-time Gauss' Theorem (eq. 5-B.8.1.1) multiplied by $(i/2)^2$ is

$$I = -2^{-2}\int d^2X d^2X^*\ (gg^*)^{1/2}\ V^{\rho}{}_{;\rho} = -2^{-2}\int d^2X d^2X^* 2 \tag{5-B.9.2.3}$$

$$= -2^{-1}\int dX^1 dX^{1*}\ (-2i)\pi(r^2 - X^1 X^{1*}) \tag{5-B.9.2.4}$$

Using the alternate form of Green's Theorem (eq. 5-B.8.4) eq. 5-B.9.2.4 becomes

$$I = i\pi \oint_C dX^{1*}\ [r^2 X^1 - (X^1)^2 X^{1*}] = i\pi[-2i\pi r^4 + i\pi r^4] = \pi^2 r^4 \tag{5-B.9.2.5}$$

The integration of the right side of the complex space-time Gauss' Theorem (eq. 5-B.8.1.1) multiplied by $(i/2)^{-2}$ is

$$J = -2^{-2}\int d\Sigma_{\rho}\ (gg^*)^{1/2}\ V^{\rho} = -2^{-2}\Big[\int dX^2 dX^{2*} \oint_C dX^{1*}\ X^1 + dX^1 dX^{1*} \oint_C dX^{2*}\ X^2\Big] \tag{5-B.9.2.6}$$

$$= -2^{-2}2(-2i)\int dX dX^*\ \pi(r^2 - XX^*) = \pi^2 r^4 \tag{5-B.9.2.7}$$

using eq. 5-B.9.1.7 and following steps similar to eq. 5-B.9.2.5 thus confirming the theorem.

5-B.10 Some Other Useful Relations for Complex Variable Integrations

Some other relations that are helpful in evaluating integrals over complex variables are:

5-B.10.1 One Complex Variable Version of Green's Theorem

If $F(z, z^*)$ is continuous and has continuous partial derivatives in some region R and on its boundary curve C then

$$\int_R dz dz^* \, \partial F / \partial z^* = -\oint_C dz \, F(z, z^*) \qquad (5\text{-B.10.1.1})$$

5-B.10.2 An Area Integral

The following is a useful integral: the integral around a closed curve C is

$$\oint_C dz \, z^* = 2iA \qquad (5\text{-B.10.2.1})$$

where A is the area enclosed by the curve C.

5-B.11 Interpretation of Curvature in 4-dimensional Complex Space-time

The interpretation of curvature in 4-dimensional complex space-time, as we have developed it, is problematic since $R_{\rho\mu\nu\sigma}$ is complex in general and thus the curvature scalar would appear to be complex. A complex curvature scalar does not lend itself to a simple, or an intuitive, interpretation. To remedy this situation we revert to the consideration of an equivalent 8-dimensional real space-time.

5-B.11.1 Transformation to Eight-dimensional Real Space-time

In sections 5-B.4 – 5-B.7 we considered an eight-dimensional complex space-time in connection with the evaluation of volume integrals. In this section we will examine a map from coordinates in 8-dimensional complex space-time (holomorphic plus anti-holomorphic coordinates) to coordinates in an 8-dimensional real space-time. To that end we introduce an extra index to distinguish our original space of four complex coordinates X^μ from their complex conjugate coordinates $X^{\mu *}$. This index, which we will call the *holomorphy index*, will use lower case Roman alphabet letters: a, b, c, … taking the values 0 and 1. We define eight dimensional complex space-time coordinates with

$$Z^{(a\mu)} = \delta^a{}_0 X^\mu + \delta^a{}_1 X^{\mu *} \qquad (5\text{-B.11.1.1})$$

Thus a = 0 corresponds to the four holomorphic space-time coordinates and a = 1 corresponds to the four anti-holomorphic space-time coordinates. We will not distinguish between superscripts and subscripts for the holomorphy index *in Kronecker delta functions* (although a generalization of our theory might introduce a local character to the "metric" of the holomorphy index.) Thus the following forms of the Kronecker delta functions are numerically equal:

$$\delta^{ab} = \delta^{ba} = \delta_{ab} = \delta^a{}_b = \delta_a{}^b$$

for a, b = 0, 1. A Kronecker delta is equal to one if a = b and zero otherwise.

We also introduce an index for eight-dimensional space-times using Cyrillic characters (reserving Greek characters for 4-dimensional indexing) to represent a pair of indices – the first index being the holomorphy index and the second index being a 4-dimensional space-time index):

$$Z^{И} = Z^{(a\mu)} \quad (5\text{-B.11.1.2a})$$

$$\partial/\partial Z^{И} = \partial/\partial X^{(b\mu)} = \delta^{b0}\,\partial/\partial X^{\mu} + \delta^{b1}\,\partial/\partial X^{\mu *} \quad (5\text{-B.11.1.2b})$$

The value of Cyrillic letter (Ж, Б, Д, И, Ч, Ш, Я, Ю, and so on) indices range from 0 through 7, with their value determined by relations of the form

$$И = 4a + \mu \quad (5\text{-B.11.1.2c})$$

with a = 0, 1 and μ = 0, 1, 2, 3.

The introduction of holomorphy indices enables us to easily define real 8-dimensional coordinates using a "holomorphy tensor" $f^{a}{}_{b}$

$$f^{a}{}_{b} = 2^{-\frac{1}{2}}(\delta^{a}{}_{0}\delta^{0}{}_{b} + \delta^{a}{}_{0}\delta^{1}{}_{b} - i\,\delta^{a}{}_{1}\delta^{0}{}_{b} + i\,\delta^{a}{}_{1}\delta^{1}{}_{b}) \quad (5\text{-B.11.1.3})$$

with inverse

$$f^{-1b}{}_{c} = 2^{-\frac{1}{2}}(\delta^{b}{}_{0}\delta^{0}{}_{c} + i\,\delta^{b}{}_{0}\delta^{1}{}_{c} + \delta^{b}{}_{1}\delta^{0}{}_{c} - i\,\delta^{b}{}_{1}\delta^{1}{}_{c}) \quad (5\text{-B.11.1.4})$$

$$f^{a}{}_{b}\, f^{-1b}{}_{c} = \delta^{a}{}_{0}\delta^{0}{}_{c} + \delta^{a}{}_{1}\delta^{1}{}_{c} \quad (5\text{-B.11.1.5})$$

We now define the basic physical relationship of our complex 8-dimensional space-time (with coordinates denoted by Z) to a *real* 8-dimensional space-time (with coordinates denoted by x). We define the real 8-dimensional coordinates as

$$x^{Я} \equiv x^{(a\mu)} = f^{a}{}_{b}\, Z^{(b\mu)} \quad (5\text{-B.11.1.6})$$

and

$$ds^2 = g_{z(a\mu)(b\nu)} dZ^{(a\mu)} dZ^{(b\nu)} = g_{x(a\mu)(b\nu)} dx^{(a\mu)} dx^{(b\nu)} = g_{x\,ЯЧ}\, dx^{Я} dx^{Ч} \quad (5\text{-B.11.1.7})$$

where $g_{z(a\mu)(b\nu)}$ is given in matrix form as g_8 in eq. 5-B.5.2.3, and in indexed form by

$$g_{z(a\mu)(b\nu)} = \delta_a^{\ 0}\delta_b^{\ 0}\, g_{\mu\nu} + \delta_a^{\ 1}\delta_b^{\ 1} g_{\mu\nu}{}^* \qquad (5\text{-B.11.1.8a})$$

with inverse:

$$g_z^{\ (a\mu)(b\nu)} = \delta_0^{\ a}\delta_0^{\ b}\, g^{\mu\nu} + \delta_1^{\ a}\delta_1^{\ b} g^{\mu\nu *} \qquad (5\text{-B.11.1.8b})$$

and satisfying (with a summation over both b and ν)

$$g_{z(a\mu)(b\nu)}\, g_z^{\ (b\nu)(c\sigma)} = \delta_a^{\ c}\delta_\mu^{\ \sigma} \qquad (5\text{-B.11.1.9})$$

Notice ds^2 is real as defined in eq. 5-B.11.1.7. In addition it yields the relation between the metric in real space-time in terms of the complex space-time metric (with summations over repeated holomorphy indices):

$$g_{x(a\mu)(b\nu)} = f^{-1c}{}_a\, f^{-1d}{}_b\, g_{z(c\mu)(d\nu)} \qquad (5\text{-B.11.1.10})$$

with inverse

$$g_x^{\ (a\mu)(b\nu)} = f^a{}_c\, f^b{}_d\, g_z^{\ (c\mu)(d\nu)} \qquad (5\text{-B.11.1.11})$$

and satisfying

$$g_{x(a\mu)(b\nu)}\, g_x^{\ (b\nu)(c\sigma)} = \delta_a^{\ c}\delta_\mu^{\ \sigma} \qquad (5\text{-B.11.1.12})$$

Some other important relations following from the preceding discussion are (with summations over repeated holomorphy indices):

$$\partial/\partial x^{(a\mu)} = f^{-1b}{}_a\, \partial/\partial Z^{(b\mu)} \qquad \partial/\partial x_{(a\mu)} = f^a{}_b\, \partial/\partial Z_{(b\mu)} \qquad (5\text{-B.11.1.13})$$

$$Z^{(a\mu)} = f^{-1a}{}_b\, x^{(b\mu)} \qquad Z_{(a\mu)} = f^b{}_a\, x_{(b\mu)} \qquad (5\text{-B.11.1.14})$$

$$\partial/\partial Z^{(a\mu)} = f^b{}_a\, \partial/\partial x^{(b\mu)} \qquad \partial/\partial Z_{(a\mu)} = f^{-1a}{}_b\, \partial/\partial x_{(b\mu)} \qquad (5\text{-B.11.1.15})$$

In general given a mixed tensor in complex 8-dimensional space-time:

$$T_z^{\ (a\alpha)(b\beta)(c\gamma)(\ldots)}{}_{(u\tau)(v\upsilon)(w\phi)(\ldots)}$$

then its real space-time equivalent is

$$T_x^{\ (a\alpha)(b\beta)(c\gamma)(\ldots)}{}_{(u\tau)(v\upsilon)(w\phi)(\ldots)} = f^a{}_{a'} f^b{}_{b'} f^c{}_{c'} \ldots f^{-1u'}{}_u f^{-1v'}{}_v f^{-1w'}{}_w \ldots T_z^{\ (a'\alpha)(b'\beta)(c'\gamma)(\ldots)}{}_{(u'\tau)(v'\upsilon)(w'\phi)(\ldots)}$$
$$(5\text{-B.11.1.16})$$

The complex space-time inverse relation is

$$T_z{}^{(a\alpha)(b\beta)(c\gamma)(\ldots)}{}_{(u\tau)(v\upsilon)(w\phi)(\ldots)} = f^{-1a}{}_{a'}f^{-1b}{}_{b'}f^{-1c}{}_{c'}\ldots f^{u'}{}_{u}f^{v'}{}_{v}f^{w'}{}_{w}\ldots T_x{}^{(a'\alpha)(b'\beta)(c'\gamma)(\ldots)}{}_{(u'\tau)(v'\upsilon)(w'\phi)(\ldots)}$$
(5-B.11.1.17)

Eq. 5-B.11.1.11 can be expanded as:

$$g_x{}^{(a\mu)(b\nu)} = (\delta^{a0}\delta^{b0} - \delta^{a1}\delta^{b1})(g^{\mu\nu} + g^{\mu\nu *})/2 - i(\delta^{a0}\delta^{b1} + \delta^{a1}\delta^{b0})(g^{\mu\nu} - g^{\mu\nu *})/2$$
(5-B.11.1.9a)

$$= (\delta^{a0}\delta^{b0} - \delta^{a1}\delta^{b1})g^{+\mu\nu} - i(\delta^{a0}\delta^{b1} + \delta^{a1}\delta^{b0})g^{-\mu\nu}$$
(5-B.11.1.9b)

which is manifestly real. Note that the 00 and 11 holomorphic diagonal components have reversed signs. Eq. 5-B.11.1.9 implies the Minkowski metric for our eight-dimensional real space is

$$\eta_x{}^{(a\mu)(b\nu)} = (\delta^{a0}\delta^{b0} - \delta^{a1}\delta^{b1})\eta^{\mu\nu}$$
(5-B.11.1.10)

so that the metric has equal numbers of positive and negative signs: +1, –1, –1, –1, –1, +1, +1, +1 resulting in equal treatment of spatial and time-like coordinates. Fig. 5-B.11.1 has a matrix-like representation of eq. 5-B.11.1.9 that enables it to be easily visualized.

Thus we have a viable mapping to a real space-time that help us to make sense of the curvature of complex space-time as we will see in the next subsection.

5-B.11.2 Calculation of the Equivalent Real Space-time Curvature tensor

The affine connection for four-dimensional complex space-time given in eq. 5-A.5.2.1 can be extended with the addition of holomorphy indexes to

$$\Gamma_z{}^{(a\sigma)}{}_{(b\lambda)(c\mu)} = \tfrac{1}{2}g_z{}^{(a\sigma)(d\nu)}\{\partial g_{z\,(c\mu)(d\nu)}/\partial Z^{(b\lambda)} + \partial g_{z\,(b\lambda)(d\nu)}/\partial Z^{(c\mu)} - \partial g_{z\,(b\lambda)(c\mu)}/\partial Z^{(d\nu)}\}$$
(5-B.11.2.1)

which has only two non-zero parts:

$$\Gamma_z{}^{(0\sigma)}{}_{(0\lambda)(0\mu)} = \Gamma^{\sigma}{}_{\lambda\mu}$$
(5-B.11.2.2)

and

$$\Gamma_z{}^{(1\sigma)}{}_{(1\lambda)(1\mu)} = \Gamma^{\sigma}{}_{\lambda\mu}{}^{*}$$
(5-B.11.2.3)

All other components are zero due to the block diagonality of $g_{z\,(a\mu)(b\nu)}$ (and its inverse) in the holomorphy indices a and b, and due to the holomorphy in X^μ of $g_{\mu\nu}$ and its inverse.

In a similar fashion we can define a generalization of the curvature tensor (eq. 5-A.5.4.5) to eight-dimensional complex space-time as we did in sections 5-B.4 through 5-B.7 as

$$R_{z(a\rho)(b\mu)(c\nu)(d\sigma)} = \tfrac{1}{2}\,[\partial^2 g_{z(a\rho)(c\nu)}/\partial Z^{(d\sigma)}\partial Z^{(b\mu)} - \partial^2 g_{z(b\mu)(c\nu)}/\partial Z^{(d\sigma)}\partial Z^{(a\rho)} -$$

$$- \partial^2 g_{z(a\rho)(d\sigma)}/\partial Z^{(c\nu)}\partial Z^{(b\mu)} + \partial^2 g_{z(b\mu)(d\sigma)}/\partial Z^{(c\nu)}\partial Z^{(a\rho)}] +$$

$$+ g_{z(e\alpha)(f\beta)}[\Gamma_z{}^{(e\alpha)}{}_{(c\nu)(a\rho)}\,\Gamma_z{}^{(f\beta)}{}_{(b\mu)(d\sigma)} - \Gamma_z{}^{(e\alpha)}{}_{(d\sigma)(a\rho)}\,\Gamma_z{}^{(f\beta)}{}_{(b\mu)(c\nu)}] \qquad (5\text{-B}.11.2.4)$$

$$\left[\begin{array}{c:c} \tfrac{1}{2}(g_{\mu\nu} + g_{\mu\nu}{}^*) & -\tfrac{1}{2}\, i(g_{\mu\nu} - g_{\mu\nu}{}^*) \\ \hdashline -\tfrac{1}{2}\, i(g_{\mu\nu} - g_{\mu\nu}{}^*) & -\tfrac{1}{2}(g_{\mu\nu} + g_{\mu\nu}{}^*) \end{array}\right]$$

Figure 5-B.11.1. A visualization of $g_{x(a^\mu)(b^\nu)}$ in eq. 5-B.11.1.9 as a 2 × 2 matrix consisting of sums and differences of the 4 × 4 metric $g_{\mu\nu}$ and its complex conjugate $g_{\mu\nu}{}^*$ viewed as matrices.

Again we note that the only non-zero components of the extended curvature tensor are

$$R_{z(0\rho)(0\mu)(0\nu)(0\sigma)} = R_{\rho\mu\nu\sigma} \qquad \text{and} \qquad R_{z(1\rho)(1\mu)(1\nu)(1\sigma)} = R_{\rho\mu\nu\sigma}{}^* \qquad (5\text{-B}.11.2.5)$$

All other components are zero due to the block diagonality of $g_{z(a\mu)(b\nu)}$ and $\Gamma_z{}^{(a\sigma)}{}_{(b\lambda)(c\mu)}$ in the holomorphy indexes, and due to the holomorphy in X^μ of $g_{\mu\nu}$ and $\Gamma^\sigma{}_{\lambda\mu}$.

We now wish to map the curvature tensor defined in eq. 5-B.11.2.4 to the real eight-dimensional space-time with coordinates x^μ. Again we use V to implement the mapping:

$$R_{x(a\rho)(b\mu)(c\nu)(d\sigma)} = f^{-1a'}{}_a f^{-1b'}{}_b f^{-1c'}{}_c f^{-1d'}{}_d\, R_{z(a'\rho)(b'\mu)(c'\nu)(d'\sigma)} \qquad (5\text{-B}.11.2.6)$$

After some algebra and the use of eq. 5-B.11.2.5 we find

$$2R_{x(a\rho)(b\mu)(c\nu)(d\sigma)} = [(\delta_a{}^0\delta_b{}^0 - \delta_a{}^1\delta_b{}^1)(\delta_c{}^0\delta_d{}^0 - \delta_c{}^1\delta_d{}^1) -$$

$$- (\delta_a{}^0\delta_b{}^1 + \delta_a{}^1\delta_b{}^0)(\delta_c{}^0\delta_d{}^1 + \delta_c{}^1\delta_d{}^0)]R^+{}_{\rho\mu\nu\sigma} +$$

$$+ \, i[(\delta_a{}^0\delta_b{}^1 + \delta_a{}^1\delta_b{}^0)(\delta_c{}^0\delta_d{}^0 - \delta_c{}^1\delta_d{}^1) +$$

$$+ \, (\delta_a{}^0\delta_b{}^0 - \delta_a{}^1\delta_b{}^1)(\delta_c{}^0\delta_d{}^1 + \delta_c{}^1\delta_d{}^0)] \, R^-{}_{\rho\mu\nu\sigma} \qquad (5\text{-B}.11.2.7)$$

where

$$R^\pm{}_{\rho\mu\nu\sigma} = (R_{\rho\mu\nu\sigma} \pm R_{\rho\mu\nu\sigma}{}^*)/2 \qquad (5\text{-B}.11.2.8)$$

The Ricci tensor in the real eight-dimensional space-time with coordinates $x^{(a\mu)}$ is

$$R_{x(a\rho)(d\sigma)} = -g_x{}^{(b\mu)(c\nu)} \, R_{x(a\rho)(b\mu)(c\nu)(d\sigma)} \qquad (5\text{-B}.11.2.9)$$

$$= -(\delta_a{}^0\delta_d{}^0 - \delta_a{}^1\delta_d{}^1)(g^{+\mu\nu} R^+{}_{\rho\mu\nu\sigma} + g^{-\mu\nu} R^-{}_{\rho\mu\nu\sigma}) \, -$$

$$- \, i(\delta_a{}^0\delta_d{}^1 + \delta_a{}^1\delta_d{}^0)(g^{-\mu\nu} R^+{}_{\rho\mu\nu\sigma} + g^{+\mu\nu} R^-{}_{\rho\mu\nu\sigma})$$

$$= -(\delta_a{}^0\delta_d{}^0 - \delta_a{}^1\delta_d{}^1)(g^{\mu\nu} R_{\rho\mu\nu\sigma} + g^{\mu\nu *} R_{\rho\mu\nu\sigma}{}^*)/2 \, -$$

$$- \, i(\delta_a{}^0\delta_d{}^1 + \delta_a{}^1\delta_d{}^0)(g^{\mu\nu} R_{\rho\mu\nu\sigma} - g^{\mu\nu *} R_{\rho\mu\nu\sigma}{}^*)/2$$

$$= (\delta_a{}^0\delta_d{}^0 - \delta_a{}^1\delta_d{}^1)(R_{\rho\sigma} + R_{\rho\sigma}{}^*)/2 \, +$$

$$+ \, i(\delta_a{}^0\delta_d{}^1 + \delta_a{}^1\delta_d{}^0)(R_{\rho\sigma} - R_{\rho\sigma}{}^*)/2$$

The curvature scalar in the real eight-dimensional space-time with coordinates x^μ is

$$R_x = g_x{}^{(a\rho)(d\sigma)} \, R_{x(a\rho)(d\sigma)} \qquad (5\text{-B}.11.2.10)$$

$$= (g^{\rho\sigma} R_{\rho\sigma} + g^{\rho\sigma *} R_{\rho\sigma}{}^*) = R + R^* \qquad (5\text{-B}.11.2.11)$$

which is manifestly real and what one might intuitively expect.

Thus we have mapped our eight-dimensional complex space-time to an eight-dimensional real space-time in which a real curvature scalar exists with a physically understandable meaning. Since complex four-dimensional space-time was defined as a holomorphic extension of real four-dimensional space-time, and since eight-dimensional complex space-time was defined as the direct sum of complex space-time and its complex conjugate space-time, it is natural to define the physical curvature of complex space-time as the real part of the curvature scalar R:

$$R_{zPhysical} = ½ (R + R^*) = \mathrm{Re}\, R \qquad (5\text{-B.11.2.12})$$

This definition of the physical curvature scalar for four-dimensional complex space-time coincides with the definition of the scalar curvature for its four-dimensional real space-time subspace – *our* physical space-time. And $R_{zPhysical}$ does take account of the impact of both the real and imaginary parts of the complex coordinates on the curvature.

5-B.11.3 Spinor Formulation of 8-dimensional Space-time

The preceding discussions of eight-dimensional real and complex space-times can be formulated using a spinor notation. The Pauli matrices σ^i have the matrix elements:

$$\sigma^{1a}{}_b = \delta^a{}_0\delta^1{}_b + \delta^a{}_1\delta^0{}_b \qquad (5\text{-B.11.3.1})$$

$$\sigma^{2a}{}_b = -\,i\,\delta^a{}_0\delta^1{}_b + i\,\delta^a{}_1\delta^0{}_b \qquad (5\text{-B.11.3.2})$$

$$\sigma^{3a}{}_b = \delta^a{}_0\delta^0{}_b - \delta^a{}_1\delta^1{}_b \qquad (5\text{-B.11.3.3})$$

where the index a labels rows and the index b labels columns. The matrix elements of the identity matrix I are:

$$I^a{}_b = \delta^a{}_0\delta^0{}_b + \delta^a{}_1\delta^1{}_b \qquad (5\text{-B.11.3.3})$$

The transformation matrix between the z and x eight-dimensional space-times can be written in terms of Pauli matrices as;

$$f = 2^{-3/2}\,[\,(1 + i)I + (1 - i)(\sigma^1 - \sigma^2 + \sigma^3)] \qquad (5\text{-B.11.3.4})$$

with inverse

$$f^{-1} = f^\dagger = 2^{-3/2}\,[\,(1 - i)I + (1 + i)(\sigma^1 - \sigma^2 + \sigma^3)] \qquad (5\text{-B.11.3.4})$$

The complex coordinates can be represented with a spinor notation:

$$Z^{(\mu)} = X^\mu\chi_0 + X^{\mu *}\chi_1 \qquad (5\text{-B.11.3.5})$$

where we indicate the matrix format with parentheses around the Minkowski indexes, and where the two component spinors are:

$$\chi_0{}^{a} = \delta^{a}{}_{0} \qquad \qquad \chi_1{}^{a} = \delta^{a}{}_{1} \qquad (5\text{-B}.11.3.6)$$

Second (and higher) order tensor quantities as well as mixed tensors lend themselves best to the indicial notation used in the preceding section. Nevertheless a spinorial formulation is suggestive in the light of supersymmetry theories and twistor theories. (See the references in section 5-A.1.)

END OF EXTRACT FROM BLAHA(2004)

6. The Equality of Inertial Mass and Gravitational Mass

From the days of Newton through Einstein[31] to the present the equality of gravitational mass and inertial mass has been a topic of interest. Mach, who played an important role, in this ongoing discussion, thought distant masses in the universe were the source of the equality. However the origin of the equality, which has been shown experimentally to very high accuracy, remained uncertain until the present work where we show the interconnection of the Extended Standard Model and Complex Gravitation via Higgs generated masses unites gravitational and inertial mass.

In chapter 5 we showed that Complex General Relativity could be formulated in a manner similar to the Extended Standard Model in which the Reality group played a part. This formulation leads to scalar particles that can be viewed as Higgs particles since the fields could be shifted by a constant without affecting the kinetic part of their dynamic equations. If a Higgs potential is present then these fields could undergo spontaneous breakdown and then have non-zero vacuu expectation values.

The decision to base gravitational Higgs fields on the Reality group transformations of Complex General Relativity was based on a desire to build a Theory of Everything. (See chapter 7.) The present Complex Gravity theory, that we have developed, directly entwines gravitation and particle physics through the Reality group. The gravitational Higgs equations then become an elegant, compact unifying feature of a Theory of Everything. The known Higgs equations of the Extended Standard Model are now intertwined with the gravitational Higgs equations, which are a consequence of Complex General Relativity (suitably extended).[32] The gravitational Higgs potential energy-momentum contribution remains to be justified but can be provisionally inserted by hand just as Higgs potentials are inserted in The Extended Standard Model.

Since fermion field masses are now sums of ElectroWeak Higgs contributions, Generation group Higgs contributions, and gravitational Higgs contributions and since the gravitational Higgs fields appearin all fermion masses the equality of inertial and gravitational mass is proven. The gravitational Higgs particles equations depend, in part, on the gravitational field by eq. 5.50 and so set the mass scale of the gravitational mass. The presence of the gravitational Higgs contributions ***for all fermions****, sets the scale of the inertial Higgs field contributions from the Extended Standard Model Higgs particles.*

Since an expression cannot mix mass scales, the gravitational mass scale must be the same as the inertial mass scale. Inertial Mass equals gravitational mass.

---

[31] For example, Einstein and Grossman in 1913 stated, "The theory herein described originates in the conviction that the proportionality between the inertial and gravitational mass of a body is an exact law of nature that must be expressed as a foundation principle of theoretical physics."

[32] This approach is further supported by the use of the Higgs mechanism to produce cosmic inflation and justify the expansion of the universe. See Guth and colleagues for discussions of Higgs induced inflation.

We have established the equality of inertial and gravitational mass at the short distance quantum level. In our view, this explanation is far more satisfying than basing the equality on a combination of large distance phenomena and quantum phenomena. As Einstein and Weyl have pointed out: all fundamental physics phenomena should be based on a local theory. Complex Gravity, as we have constructed it, combined with the Extended Standard Model furnishes a completely local basic Theory of Everything.

7. SU(3)⊗SU(2)⊗U(1)⊗SU(2)⊗U(1)⊗U(4) Theory of Everything

There have been many attempts to develop a Theory of Everything (TOE): in modern times from at least the 1900s; in Classical Times from at least the period of the Pre-Socratic Philosophers. The author, as well as many other physicists, has attempted to develop a TOE in the past forty years – hopeful that at last the fundamental interactions of Nature have been understood. Perhaps the most popular attempt has been in the form of SuperString theories. However theoretic difficulties, and the experimental lack of any significant supporting data, have led to a dimunition of interest.

Theoretically, the most basic question that is not addressed by SuperStrings is "Why SuperStrings?" There is no fundamental justification for string constructs. And one can also question whether there is a still more fundamental theory – strings are not fundamental simple constructs.

This author's efforts over the past fifteen years have been based on Quantum Field Theory – a theory with an outstanding record of success in predicting perturbative corrections to fundamental constants in Quantum Electrodynamics with unrivaled precision.[33]

The author came to the conclusion that "All we know with certainty is that Logic must hold in Physics."[34] Based on that conclusion the author constructed a Quantum Field Theoretic Extended Standard Model whose features are outlined in chapter 1, and a new theory of Complex General Relativity described in chapter 5.

The satisfactory unification of these theories constitutes a Theory of Everything.

7.1 What is an Acceptable Theory of Everything?

It seems necessary to consider what constitutes a TOE. One cannot simply shoehorn a theory of elementary particles (generalized to curved space-time) together with gravitation with no other connection and call the result a Theory of Everything. Rather, a TOE should involve a deeper union of spirit and theoretic constructs that somehow truly unites them. Here we are reminded of Pauli's great quip: "One should not unite things, which God has put asunder." So one expects many hooks between the parts of a TOE.

---

[33] See the papers of Professor Kinoshita, and colleagues, for their extremely precise calculations.

[34] The author also showed that Gödel's Theorem had been misunderstood and was not a failure of Logic or logical reasoning. See Blaha (2011c).

7.2 Components of a Theory of Everthing

We believe that the Extended Standard Model and the new Theory of Complex General Relativity form a Theory of Everything. The components are described in the following sections.

7.3 Fundamental Approach

The fundamental approach to the development of the TOE is based on Asynchronous Logic, which is known to be required to support parallel processes – an essential requirement for a non-trivial theory. From its 4-valued logic we obtain the dimensionality of space-time and then proceed to develop the physics of flat space-time (The Extended Standard Model) and curved space-time (Complex General Relativity).

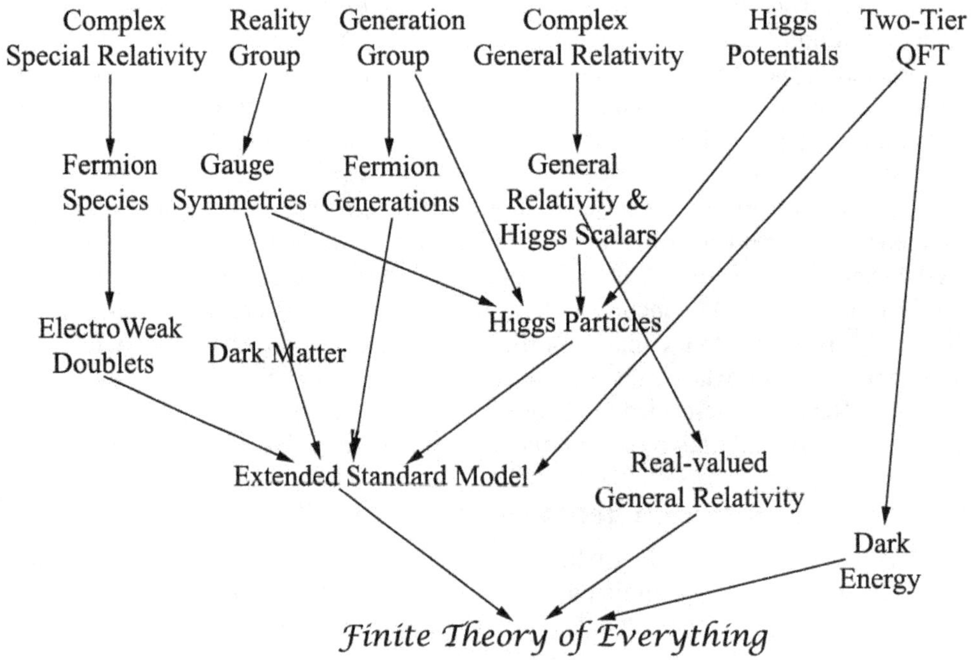

Figure 7.1. A diagram of the connections between the various parts of the Theory of Everything.

7.4 Impact of Reality Group on the Extended Standard Model and Complex General Relativity

Since space-time (both flat and curved), while complex, can only be measured with instruments that yield real values, we introduce the Reality group for both theories. In the case of The Extended Standard Model we obtain tachyons, parity violation, four species of fermions, SU(3)⊗SU(2)⊗U(1)⊗SU(2)⊗U(1) symmetry (broken), gauge invariance, and Higgs particles with non-zero vacuum expectation values. In the case of Complex General Relativity we find the Einstein dynamic equations with gauge field aspects, and a set of 16 scalar fields that can support a Higgs Mechanism to generate part of the masses of Extended Standard Model particles.

7.5 A Unified Theory - TOE

The preceding subsections show the unity of the parts – a common origin in geometry, a commun mapping to real coordinates with major consequences, and a common unification of the mass structure in the particle dynamic equations. Fig. 7.1 shows the connectedness of our TOE.

The proof of the pudding: at last we have a local chain of causation for the equality of inertial mass and gravitationa mass: inertial mass being the mass of particles, gravitational mass being an artifact of gravity – particularly of the gravitational Higgs particles that have gravitational interaction terms and contribute to the masses of elementary particles.

7.6 What Next after the TOE?

If our TOE is the correct Theory of Everything and is so proven by experiments, then the question arises of what Elementary Particle Physicists should do next. It appears there are several lines of further dvelopment needed:

1. Going to much higher energies to confirm the possible existence of new forces and types of particles beyond the eight known species of fermions and the known gauge bosons.

2. Investigating new two and three quark hadrons to further understand the elusive nature of the strong SU(3) interaction.

3. Investigating new multi-quark particles to determine their properties, to further understand the strong interaction, and hopefully to find an "island of stability" in which multi-quark particles have significantly longer lifetimes. This possibility is similar to the studies of atomic nuclei which apparently has an "island of stability." Multi-quark physics has the flavor of nuclear physics with its shell model and liquid drop model. This approach is of interest because there is evidence that pentaquarks consist of a two-quark part and a three-quark part – perhaps the first evidence of a deeper shell-like structure.

4. The further study of particle interactions to better understand the nature of particle interactions mindful of the difficulty of finding/understanding solutions for the Yang-Mills gauge fields that constitute particle interactions. In a sense this study is similar to chemistry – we would know the basic theory but seek to understand the convoluted behavior of quark and gauge boson aggregates.

5. The discovery of quark-gluon plasmas in a liquid-like state at Brookhaven and CERN is of great interest. The study of this phenomenon at much higher energies and ion beam intensities should be pursued. It opens the possibility of practical applications such as advanced fusion reactors and ion drives for spaceships and starships.

6. A deeper understanding of the TOE would be desirable – particularly of the nature and justification of Higgs particles. The fermions "come from" Asynchronous Logic and Complex Special Relativity; the gauge boson symmetries and fields "come from" the Reality group; the fermion generations "come from" the Generation group. From where do the Higgs particles come? What deeper principle?

Particle Physics is still a child of the 20th century. What shall it become?

8. Features of the Megaverse

In volume I we discussed the possibility of a space outside our universe that we called the *Megaverse* after William James.[35] We suggested that there are good reasons to believe that it exists due to Quantum Gravity, and the nature of the Higgs Mechanism of The Extended Standard Model and Complex Gravitation. In this chapter we summarize some of the features presented in volume I, and add new thoughts on the Megaverse and its features.

Briefly put, the reasons for believing our universe resides in a larger space containing other universes are:

1. The existence of other universes in the Megaverse provides "Observers" that make Quantum Gravity within our universe consistent with a Copenhagen interpretation.

2. The existence of other universes in the Megaverse provides a "clock" that sets the time in our universe.

Chapter 23 of volume I discusses these topics in more detail.

8.1 The Universes within the Megaverse

The first issue is the dimensionality of universes within the Megaverse. In volume I, and in earlier books, we showed that the dimensionality of our universe is *four* based on the 4-valuedness of Asynchronous Logic. Lower dimension universes would have a much more limited range of phenomena as past field theory studies have shown. These "half-baked" universes might exist but their inclusion in our studies would not yield new insights.

Larger dimension universes would have a larger Reality group and so more interactions (more symmeties and gauge fields) as well as more fermion particle species and generations. Larger dimension universes cannot be ruled out but they can be viewed as not "economical" since they give added complexity without novel new features.

We therefore confine our studies to a Megaverse populated with 4-dimensional universes.

8.2 Types of Universe Particles

The four types of *fermion* universe particle spin 127/2 equations of are:

Dirac-type universe
Tachyonic-type universe

---

[35] The Megaverse is not the Multiverse of Everitt and others but an actual space within which our universe resides together with other universes.

Dirac-type universe with a complex 15-momentum[36]
Tachyonic universe with a complex 15-momentum

Tachyonic universes are moving with a 15-velocity whose magnitude is greater than c – the same speed of light as within our universe. The Wheeler-DeWitt equations have a negative mass squared term indicating a tachyonic particle interpretation. (See subsection 23.16.1 in volume I.)

The first consideration in the discussion of fermionic universes (reminiscent of the discussions of spin in the 1920's) is the interpretation of spin 127/2 states. We suggest that the upper (128) components (64 "spin up" and 64 "spin down") of a universe wave function represent a universe with an excess number of baryons. The lower (128) components lead to an anti-universe where there is an excess of anti-baryons. These associations are analogous to the interpretations of the Dirac electron wave function.[37]

The universe particle "spin up" and "spin down" states are distinguished by their interactions with Megaverse gauge fields in a manner analogous to quantum electrodynamics.

Spin 127/2 universe particles can also have "handedness." The physical meaning of universe handedness is an interesting issue. When we consider our universe we see left-handedness in the weak interactions of elementary particles. In addition it appears that organic molecules overwhelmingly favor left-handedness on earth although right-handed molecules exist in outer space and can be created in the laboratory. Right-handed molecules transform into left-handed molecules in watery media through electromagnetic effects.

Why nature favors left-handedness is an open question. It has given rise to speculations that gravitation, especially quantum gravitons, may be left-handed. The European Space Agency's Planck telescope will study polarization effects in the cosmos and may show that gravitons starting from the beginning of the universe, and magnified by inflation in the universe's expansion, are left-handed.

If handedness of gravitation is verified experimentally, then our theory of left-handed/right-handed universe particles would be supported. *Our universe would then be tachyonic within the Megaverse and probably left-handed. We, in this universe, would of course not necessarily know of the velocity of the universe in the Megaverse.*

Bosonic universes may also exist. A bosonic universe could be viewed as a universe with equal numbers of baryons and anti-baryons.

8.3 Megaverse Characteristics

We now have a view of the general nature of universes in the Megaverse. The next question we examine is the general nature of the Megaverse itself. *Ockham's Razor implies that complex coordinates universes require embedding in a complex Megaverse.* If the Megaverse

---

[36] We will show the Megaverse has 16 dimensions in section 8.3. Since the Megaverse has 16 complex dimensions, spin 127/2 universe particles have 256 components.

[37] It is known that phenomena in our universe tend to be left-handed. If this feature of our universe's phenomena is also a property of the universe itself, then, since handedness is an attribute of spin, the treatment of a universe as having spin is not unreasonable.

had real-valued coordinates an embedding of our universe, and other universes, in a flat Megaverse would require the Megaverse to have a minimum of ten dimensions. This type of embedding has been known since the early 20th century. The embeddings then were usually within a real-valued 10-dimensional flat space since a 4-dimensional real-valued metric $g_{\mu\nu}$ has 10 independent components.

We assumed in volume I the Megaverse has 16 complex dimensions in order to have a smooth transition between the Reality group of a 4-dimensional universe and the Reality group of the Megaverse. Section 9.4 below describes the transition between Reality groups.

In volume I we introduced quantum interactions between universes that can also probe the contents of universes. We were guided by an analogy with electromagnetic probes of protons: for low energies probes treat a proton as an elementary particle with a spatial extension; for very high energies probes "penetrate" a proton revealing its quark-parton constituents. Thus a Megaverse quantum "observer" can use interactions, such as the baryonic field, to observe the structure within universes. Chapter 24 of volume I considered baryonic and Dark baryonic probes of universes.

8.4 Summary of Features of the Megaverse

There are a number of features of the Megaverse that follow from previous discussions here and in volume I:

- The Megaverse has non-zero gravitation and curvature outside of universes, not due to the mass of a universe but due to the energy of the Baryonic field, Dark baryonic field, Reality group gauge fields, and the Y^i(y field), and quantized universe particles.

- There are baryonic and Dark baryonic long range gauge fields in universes and the Megaverse. They exert forces between a universe and the baryons within the same universe, and between a universe and the baryons within another universe. At low energies the forces are between universes. At high energies baryons in different universes interact directly.

- We can define a mass for a universe (see below) that is time dependent and proportional to the area of the universe in a manner analogous to black holes.

- Universes can be treated as second quantized particles with either spin 127/2 or or spin zero – universe particles. Spin 127/2 universe particles appear implied by the "handedness" that we see in our universe. Higher spin universes are possible but not as probable as low spin universes in the author's view.

- Each universe occupies a region of the Megaverse that is a closed set in Megaverse coordinates, and a partially open, and partially closed, set in the curved coordinates of the universe. See section 9.7 below.

- Within the boundaries of a universe one can use either Megaverse coordinates or the curved coordinates of the universe.

- The baryon, Dark baryon and Reality group gauge fields can create universe-antiuniverse pairs. Our universe may have been created through such a Big Bang process.

- Due to the size of universes, the creation or interaction of universes via the baryonic fields will have form factors or structure constants analogous to similar features in the electromagnetic behavior of hadrons.

- It is possible that universes can be tachyonic. This is explicitly seen in chapter 24 of volume I in the solutions of the Wheeler-DeWitt equation.

- We assume that other universes have the same physics as our universe with the possible difference that they may have different coupling constants and particle masses.

s.

9. The Megaverse Theory of Everything

9.1 Matter and Fields in the Megaverse

The matter and fields in the Megaverse consists of

- universe particles[38] whose name we will abbreviate to *Uons;*
- baryonic and Dark baryonic gauge fields;
- a gravitational field whose sources are the mass-energy of Megaverse matter and energy, but not the matter within universes except through their identity as Uons;
- ordinary particles at low density, and gauge fields of the same sort as within universes;
- the other gauge fields of the Extended Standard Model;
- Higgs particle fields; and
- 256 gauge fields corresponding to the Reality group of the 16-dimensional Megaverse.

The following sections consider each of these types of particles and fields.

9.2 Universe Particles – Uons

Uons have been described in some detail in volume I. These particles consist of entire 4-dimensional universes treated as quantum field theory entities. They are an extension of the development of Quantum Gravity within our universe, and particularly of, the Wheeler-DeWitt equation, which, in a sense, is the most advanced notion embodying Quantum Gravity.

With Black Hole theory providing guidance, we attribute baryon and Dark baryon numbers and associated gauge fields, an area within 16-dimensional space that is interpreted as a mass (like Black Holes), zero charge, and a spin – all to universes. These features appear to be the sole characteristics of a universe based on "no-hair" theorems. In volume I we discussed Uons with spins of zero and 127/2 (the minimal spin of a fermion in 16 dimensions).

We attributed a mass M to Uons depending on their area but not depending on the mass-energy within them since mass-energy is not well defined within a universe:

$$M = \kappa A/8\pi \qquad (9.1)$$

where A is the area of the universe. This relation is a result of Black Hole theory.

[38] Universe particles were described in detail in volume I. We choose to abbreviate their name to Uons for convenience.

We defined Dirac-like equations for Uons modeling them on Dirac-like equations. We suggested that Uons interact with baryon and Dark baryon fields due to their baryonic "charges", the Megaverse gravitational field external to universes, Megaverse Higgs particles external to universes, and the 256 gauge fields of the Megaverse Reality group. They do not directly interact with ordinary (or Dark) matter in the Megaverse. They do not participate in the interactions of the Extended Standard Model operative within universes although they do interact with a Megaverse form of the Extended Standard Model, the *Megaverse Standard Model* described later. This interaction is limited by the fact that Uons are scalars under SU(3) and SU(2)⊗U(1)⊗SU(2)⊗U(1) – thus the lack of Standard Model interactions between Uons and matter scattered in the Megaverse.

Uons can be created and destroyed through a mechanism similar to the creation of electron-positron pairs by photons. In the case of Uons baryonic (and Dark baryonic) particles that we called plancktons in volume I can create and annihilate Uon-antiUon pairs (volume I).[39] Since universes of small size can exist such as our universe at the instant of the Big Bang the creation and annihilation of Uon pairs could take place at comparatively low energies. After a Uon is created it can grow in size (area) and mass with time – an expansion of a universe. It could also contract with time. Volume I describes the expansion/contraction scenario in detail with Uon propagators.

Given a matter density in the Megaverse it is reasonable to expect Megaverse matter may enter into a universe. This process, which if it exists should be a continuing phenomena, could effectively resemble the random "creation" of matter in a universe – a process that could partly fuel the expansion of the universe. "Continuous creation of matter" theories were a subject of discussion in the 1950s – 1970s – particularly by Hoyle and Narlikar.

An interesting feature of our universe is its lopsidedness with significantly more matter and energy on one side – a characteristic that one would expect of an accelerated liquid on earth. The Uon model of universes within the Megaverse could explain this phenomena if our universe had been accelerated within the Megaverse at some early point in its history – perhaps just after creation in a pair creation event. In that time period, the matter and energy distribution within the universe would become lopsided. The lopsided character would then persist as the universe expanded.

9.3 Uon-AntiUon Creation

It can be expected that the creation of a Uon-antiUon pair would create a sort of "mirror-opposite" pair of universes just as electron-positron creation produces creates oppositely charged particles.

There are two interesting possible mirror opposite effects: 1) the Uon may contain a preponderance of particles and the antiUon might contain a corresponding preponderance of antiparticles; 2) the Uon might be left-handed and correspondingly the antiUon might be right-handed. These characteristics could explain two striking aspects of our universe: it appears to

---

[39] Pair production of universes can take place in the vicinity of a source of baryon fields such as an existing Uon through a planckton transitioning to a Uon pair.

have a preponderance of particles;[40] and the universe appears to be left-handed. We see the left-handedness in organic phenomena on earth. We see a similar left-handedness in the cosmos in interstellar clouds and nebulae. The concept of pair creation of universes appears intellectually more satisfying to the author than other proposals that might be entertained. It also provides an answer to the repeated question: What existed before our universe was created? However it does leave the question of the origin of the Megaverse and the "original" universes within it unanswered.

9.4 Megaverse Standard Model

The Extended Standard Model that we derived in volume I (and earlier) holds within our universe. Since its basis is Asynchronous Logic and Complex Special Relativity, and since it predicts a 4-dimensional universe, we shall assume all universes embody the Extended Standard Model.

The Extended Standard Model of a universe does not extend beyond the boundaries of a universe. Universes confine Standard Model properties except for total charge and spin. Universes appear to be uncharged. Thus there is no source for a universe's Extended Standard Model features outside a universe. Uons are Standard Model scalars.

The Megaverse itself has a form of the Extended Standard Model, The Megaverse Standard Model, that differs from our universe's Extended Standard Model in its particles, its interactions, and its values for coupling constants and masses.

The key difference is in the Reality group. The Reality group of the Extended Standard Model of our universe is R = SU(3)⊗SU(2)⊗U(1)⊗SU(2)⊗U(1) with 16 generators. The Reality group of complex 16-dimensional space, requires $16^2 = 256$ generators. While it is tempting to view the 16-dimensional Reality group as U(16) since it certainly maps all points in complex 16-dimensional space to real valued coordinates, it's choice would lead to an extensive, complicated fermion spectrum that is not present in our universe.

It is reasonable to require continuity in the fermion matter spectrum between our universe and the Megaverse so that the same set of particles would exist in both. Then particles could travel between our universe and the Megaverse. To achieve this goal in the simplest manner we will require the 256 generators of the 16-dimensional Reality group to be the product of sixteen commuting copies, denoted R_i, of the 4-dimensional Reality group R:

$$R_{16} = \prod_{i=1}^{16} \otimes R_i \tag{9.2}$$

The resulting set of 256 subgroup generators should algebraically span the set of 256 generators of U(16) just as the 16 generators of R span the set of 16 generators of U(16). Then the covariant derivative terms of the Megaverse Standard Model will have each SU(3) and each SU(2) gauge field replaced with a sum over the fields within the set of R_i:

[40] The distinction between particles and antipartcles is a matter of definition.

$$\sum_{i=1}^{16} g_i \mathbf{G}_i \cdot \mathbf{U}_{i\mu} \qquad (9.3)$$

and similarly for U(1) fields:

$$\sum_{i=1}^{16} g'_i B'_\mu \qquad (9.4)$$

The kinetic lagrangian terms for the sixteen copies are simply the sum of sixteen lagrangian terms copies similar to the corresponding lagrangian term in the Extended Standard Model of our universe. The Higgs particles that give masses to fermions and gauge particles are similar to those of the Extended Standard Model. They amount to sixteen copies of the Extended Standard Model terms of our universe – possibly with different vacuum expectation values and different coupling constants.[41]

The Generation group part of The Extended Standard Model carries over without change except for the requirement that the fields – indeed all fields of the Megaverse Standard Model – are 16-dimensional fields as shown in volume I.

Thus the Megaverse Standard Model has a remarkable similarity in form to the Extended Standard Model of our universe although the coupling constants and masses may differ.

9.5 Megaverse Normal and Dark Matter

The Megaverse should have the equivalents of normal and Dark matter at low density[42] found in our universe: leptons, fermions, gauge particles, Higgs particles and so on since the Megaverse Standard Model, as we have defined it, has almost the same form as our Extended Standard Model.

However, the masses and coupling constants would possibly differ. The result would be a very different form of matter that is impossible to ascertain at present. This circumstance would make travel by humans outside the universe problematic.

It is also possible that the Extended Standard Model is the same in form as the Megaverse Standard Model. No one knows how the constants of the Standard Model are determined. It is possible that the differences between being present in a universe and being present in the Megaverse do not affect the value of constants. In this happy case it would be possible to travel between a universe and the Megaverse without the difficulties of changes in the properties of matter.

---

[41] It is possible that the Megaverse coupling constants and masses conspire to yield the corresponding Extended Standard Model quantities. If so, then the passage of matter between a universe and the Megaverse would have no difficulties of continuity and boundary conditions at the transition point. If not, then a particle traveling between these regions would undergo a transformation. This possible phenomenon would be resolvable by matching boundary values. It would create a problem for the passage of life between our universe and the Megaverse.

[42] The density would presumably much lower than the average density of matter between galaxies in our universe.

9.6 Megaverse Gravitational Field

The Megaverse gravitational field exists in a 16-dimensional complex space. We can perform Reality group transformations on the coordinates to introduce a time dimension since it is required for dynamical evolution, and 15 initially[43] *real-valued spatial coordinates. Then it is straightforward to develop the Einstein dynamical equations for 16 dimensions since there is minimal dependence on the dimension in the manipulations leading to it. Einstein equations have the form:*

$$R_{\mu\nu}(x'') - \beta g_{\mu\nu} R(x'') = -8\pi G T_{\mu\nu} \qquad (9.5)$$

where β and G are constants.

The energy-momentum tensor $T_{\mu\nu}$ in the Megaverse has contributions from Uons, matter, gauge fields and Higgs fields. The mass-energy within universes does not directly contribute to $T_{\mu\nu}$.

A reasonable first view of the effect of the gravitational field is would be to think of the Megaverse as containing Uons in motion in a manner reminiscent of galaxies within our universe. The scale is different but there is a conceptual similarity. Matter would also be moving in response to gravitation.

Can one view the Megaverse as a "super-universe?" Possibly, but one would have to ask whether such a view has significant consequences. The question of whether the Megaverse is closed or not is also of intellectual interest. However, due to the remoteness of possible experimental study for the foreseeable future it appears the main focus of interest should be the interface between our universe and the Megaverse.

9.7 Megaverse-Universe Interface

Every point in our universe has 16 Megaverse coordinates in addition to our 4-dimensional space-time coordinates. If we examine the universe near its presumed boundary with the Megaverse at infinity (with one or more infinite-valued space-time coordinates), then we see the boundary of our universe at those points is open (like an open set) from the viewpoint of our universe, and closed from the viewpoint of the Megaverse.

At finite-valued coordinates in our universe, the boundary is a closed set from the viewpoint of both our universe and the Megaverse. We could call the boundary at these points a common boundary.

We can define continuity conditions for the transition of fields and other quantities from our universe to the Megaverse. The continuity conditions are to some extent illustrated in continuity conditions on the light cone in our universe.[44] We shall not consider this further in

---

[43] Subsequently we would complexify coordinates using the Higgs particles mechanism of chapter 5.

[44] Blaha, S., "Relativistic Bound State Models with Free Constituent Motion at Space-like Distances," Phys. Rev. **D12**, 3921 (1975).

the present work. Instead we shall consider the practical issue of whether we can, or cannot, cross the boundary, and, if so, how we may cross it.

*The boundary between our universe and the Megaverse is both elusive and existent at every point of our universe. Each point in our space-time has a Megaverse location – Megaverse coordinates. So it would appear that we could just slip through to the Megaverse with a small "wiggle." However, we are held back by the mighty constraint of conservation of momentum. We can only have momentum with a spatial direction within our universe. T**here is no way to throw in a direction not within our universe. All macroscopic forces lie within our universe with two possible exceptions – baryon and Dark baryon forces (discussed later).***

One might suggest that we could exit the universe in a quantum way: let a particle decay into a pair of particles and hope that one particle will move in a direction outside the universe, allbeit with perhaps very low probability. A quantum leak into the Megaverse? Not possible. For we define quantum theory in the coordinates of our universe. So the probability of quantum escape is eliminated *ab initio*.

Thus we seem unequivocally bound to our universe despite every point of our universe being infinitesimally close to a Megaverse point.

9.8 Entry and Exit from the Megaverse

There is one possible mechanism for exiting our universe, which we have described in a series of books beginning with Blaha (2014b) on uniship travel into the Megaverse. In volume I we described the Baryonic force which seems to exist based on the discrepancies in measurements of the gravitational constant G. Starships (called *uniships*) should experience a large baryonic force near large bodies such as neutron stars. The baryonic force should have a 16-dimensional Megaverse formulation as shown in volume I since universes have baryonic "charge."

On this basis we proposed a starship design with an flexible array of 16 thrust tubes arranged in a circular pattern around the ship that would each experience varying amounts of the Coulomb baryonic force that would swivel the tubes in 16 different directions in the Megaverse achieving Megaverse entry. We suggested that a starship slingshot around a neutron star accomplish this maneuver. Reentry into the universe would amount to positioning the uniship within the universe with a direction of motion within the universe.

Maneuvering into a universe is a simpler matter in the Megaverse since all directions in the Megaverse are accessible unlike the situation within the interior of universes.

9.9 Quantum Communication between a Universe Location and a Megaverse Location

Quantum communication is a rapidly developing field. At this point in time it has only been demonstrated for short ranges. Eventually it should mature as a long distance communication device such as that described in Blaha (2914c). Its great point is its instantaneous nature. Instantaneous communication at very great distances becomes feasible.

Thus one can hope to communicate across the universe. One can also hope for instantaneous communication with ships in the Megaverse. Sending a uniship into the Megaverse should not interfere with quantum communication using a detector at home and another that traversed into the Megaverse on a uniship. The difference in the number of dimensions should not affect the communication between the quantum devices. We expect that quantum states straddling the universe-Megaverse border should not go "out of sync."

Therefore quantum communication offers the possibility of instantaneous communications in our universe, and with uniships/colonies in the Megaverse and other Megaverse universes.

10. Concluding Comments

This volume concludes the derivation of Blaha's Extended Standard Model as far as it is known. The features of experimentally known Standard Model are derived within it and volume I. Dark Matter and Dark Energy are represented within Blaha's extended theory – Dark Matter as a result of an additional SU(2)⊗U(1) factor in the Reality group R (consisting of R = SU(3)⊗SU(2)⊗U(1)⊗SU(2)⊗U(1)) and Dark Energy arising from the Y^{μ} gauge field in the Two-Tier Quantum Field Theory formalism used in Blaha's theory to eliminate perturbation theory divergences. The Y^{μ} field also surpresses infinities in the state of the universe at the time of the Big Bang.

The predictions of Blaha's theory have been detailed in volume I. It is now time for experimental studies to confirm (or deny) them. As the energy of accelerators increases it is possible that new phenomena will arise. Then extensions of the theory that we propose would need to be considered. The source of these extensions, if they are found, remains to be determined. The strong basis of our theory in the geometry of complex space-time, and the lack of further geometrical structure to extend the theory, suggests that our Extended Standard Model is complete.

Perhaps the most exciting possibility for theoretical advance is the development of a detailed understanding of quark-gluon plasmas (as seen in high energy nuclei collisions). It is of importance not only for the light it will shed on color confinement but also because it offers the possibility of faster-than-light ion propulsion that can expeditiously take Man to the stars, to other galaxies, and perhaps out of our universe to the Megaverse as detailed in volume I and the author's earlier books.

REFERENCES

Akhiezer, N. I., Frink, A. H. (tr), 1962, *The Calculus of Variations* (Blaisdell Publishing, New York, 1962).

Bjorken, J. D., Drell, S. D., 1964, *Relativistic Quantum Mechanics* (McGraw-Hill, New York, 1965).

Bjorken, J. D., Drell, S. D., 1965, *Relativistic Quantum Fields* (McGraw-Hill, New York, 1965).

Blaha, S., 1998, *Cosmos and Consciousness* (Pingree-Hill Publishing, Auburn, NH, 1998).

_______, 2002, *A Finite Unified Quantum Field Theory of the Elementary Particle Standard Model and Quantum Gravity Based on New Quantum Dimensions™ & a New Paradigm in the Calculus of Variations* (Pingree-Hill Publishing, Auburn, NH, 2002).

_______, 2003, *A Finite Unified Quantum Field Theory of the Elementary Particle Standard Model and Quantum Gravity Based on New Quantum Dimensions™ and a New Paradigm in the Calculus of Variations* (Pingree-Hill Publishing, Auburn, NH, 2003).

_______, 2004, *Quantum Big Bang Cosmology: Complex Space-time General Relativity, Quantum Coordinates™ Dodecahedral Universe, Inflation, and New Spin 0, ½, 1 & 2 Tachyons & Imagyons* (Pingree-Hill Publishing, Auburn, NH, 2004).

______, *2005a, Quantum Theory of the Third Kind: A New Type of Divergence-free Quantum Field Theory Supporting a Unified Standard Model of Elementary Particles and Quantum Gravity based on a New Method in the Calculus of Variations* (Pingree-Hill Publishing, Auburn, NH, 2005).

______, 2005b, *The Metatheory of Physics Theories, and the Theory of Everything as a Quantum Computer Language* (Pingree-Hill Publishing, Auburn, NH, 2005).

______, 2005c, *The Equivalence of Elementary Particle Theories and Computer Languages: Quantum Computers, Turing Machines, Standard Model, Superstring Theory, and a Proof that Gödel's Theorem Implies Nature Must Be Quantum* (Pingree-Hill Publishing, Auburn, NH, 2005).

______, 2006a, *The Foundation of the Forces of Nature* (Pingree-Hill Publishing, Auburn, NH, 2006).

______, 2006b, *A Derivation of ElectroWeak Theory based on an Extension of Special Relativity; Black Hole Tachyons; & Tachyons of Any Spin.* (Pingree-Hill Publishing, Auburn, NH, 2006).

______, 2007a, *Physics Beyond the Light Barrier: The Source of Parity Violation, Tachyons, and A Derivation of Standard Model Features* (Pingree-Hill Publishing, Auburn, NH, 2007).

______, 2007b, *The Origin of the Standard Model: The Genesis of Four Quark and Lepton Species, Parity Violation, the ElectroWeak Sector, Color SU(3), Three Visible Generations of Fermions, and One Generation of Dark Matter with Dark Energy* (Pingree-Hill Publishing, Auburn, NH, 2007).

______, *2008a, A Direct Derivation of the Form of the Standard Model From GL(16) (Pingree-Hill Publishing, Auburn, NH, 2008).*

______, 2008b, *A Complete Derivation of the Form of the Standard Model With a New Method to Generate Particle Masses Second Edition* (Pingree-Hill Publishing, Auburn, NH, 2008)

______, 2009, *The Algebra of Thought & Reality: The Mathematical Basis for Plato's Theory of Ideas, and Reality Extended to Include A Priori Observers and Space-Time Second Edition* (Pingree-Hill Publishing, Auburn, NH, 2009).

______, 2010a, *Operator Metaphysics: A New Metaphysics Based on a New Operator Logic and a New Quantum Operator Logic that Lead to a Mathematical Basis for Plato's Theory of Ideas and Reality* (Pingree-Hill Publishing, Auburn, NH, 2010).

______, 2010b, *The Standard Model's Form Derived from Operator Logic, Superluminal Transformations and GL(16)* (Pingree-Hill Publishing, Auburn, NH, 2010).

______, 2011a, *21st Century Natural Philosophy Of Ultimate Physical Reality* (McMann-Fisher Publishing, Auburn, NH, 2011).

______, 2011b, *All the Universe! Faster Than Light Tachyon Quark Starships & Particle Accelerators with the LHC as a Prototype Starship Drive Scientific Edition* (Pingree-Hill Publishing, Auburn, NH, 2011).

______, 2011c, *From Asynchronous Logic to The Standard Model to Superflight to the Stars* (Blaha Research, Auburn, NH, 2011).

______, 2012a, *From Asynchronous Logic to The Standard Model to Superflight to the Stars volume 2: Superluminal CP and CPT, U(4) Complex General Relativity and The Standard Model, Complex Vierbein General Relativity, Kinetic Theory, Thermodynamics* (Blaha Research, Auburn, NH, 2012).

______, 2012b, *Standard Model Symmetries, And Four And Sixteen Dimension Complex Relativity; The Origin Of Higgs Mass Terms* (Blaha Reasearch, Auburn, NH, 2012).

______, 2013a, *Multi-Stage Space Guns, Micro-Pulse Nuclear Rockets, and Faster-Than-Light Quark-Gluon Ion Drive Starships* (Blaha Research, Auburn, NH, 2013).

______, 2013b, *The Bridge to Dark Matter; A New Sister Universe; Dark Energy; Inflatons; Quantum Big Bang; Superluminal Physics; An Extended Standard Model Based on Geometry* (Blaha Reasearch, Auburn, NH, 2013).

______, 2014a, *Universes and Multiverses: From a New Standard Model to a Physical Multiverse; The Big Bang; Our Sister Universe's Wormhole; Origin of the Cosmological Constant, Spatial Asymmetry of the Universe, and its Web of Galaxies; A Baryonic Field between Universes and Particles; Flatverse Extended Wheeler-DeWitt Equation* (Blaha Reasearch, Auburn, NH, 2014).

______, 2014b, *All the Multiverse! Starships Exploring the Endless Universes of the Cosmos Using the Baryonic Force* (Blaha Research, Auburn, NH, 2014).

______, 2014c, *All the Multiverse! II Between Multiverse Universes: Quantum Entanglement Explained by the Multiverse Coherent Baryonic Radiation Devices – PHASERs Neutron Star Multiverse Slingshot Dynamics Spiritual and UFO Events, and the Multiverse Microscopic Entry into the Multiverse* (Blaha Research, Auburn, NH, 2014).

Eddington, A. S., 1952, *The Mathematical Theory of Relativity* (Cambridge University Press, Cambridge, U.K., 1952).

Fant, Karl M., 2005, *Logically Determined Design: Clockless System Design With NULL Convention Logic* (John Wiley and Sons, Hoboken, NJ, 2005).

Gelfand, I. M., Fomin, S. V., Silverman, R. A. (tr), 2000, *Calculus of Variations* (Dover Publications, Mineola, NY, 2000).

Giaquinta, M., Modica, G., Souchek, J., 1998, *Cartesian Coordinates in the Calculus of Variations* Volumes I and II (Springer-Verlag, New York, 1998).

Giaquinta, M., Hildebrandt, S., 1996, *Calculus of Variations* Volumes I and II (Springer-Verlag, New York, 1996).

Gradshteyn, I. S. and Ryzhik, I. M., 1965, *Table of Integrals, Series, and Products* (Academic Press, New York, 1965).

Heitler, W., 1954, *The Quantum Theory of Radiation* (Claendon Press, Oxford, UK, 1954).

Huang, Kerson, 1992, *Quarks, Leptons & Gauge Fields 2nd Edition* (World Scientific Publishing Company, Singapore, 1992).

Jost, J., Li-Jost, X., 1998, *Calculus of Variations* (Cambridge University Press, New York, 1998).

Rescher, N., 1967, *The Philosophy of Leibniz* (Prentice-Hall, Englewood Cliffs, NJ, 1967).

Sagan, H., 1993, *Introduction to the Calculus of Variations* (Dover Publications, Mineola, NY, 1993).

Sakurai, J. J., 1964, *Invariance Principles and Elementary Particles* (Princeton University Press, Princeton, NJ, 1964).

Streater, R. F. and Wightman, A. S., 2000, *PCT, Spin, Statistics, and All That* (Princeton University Press, Princeton, NJ 2000).

Weinberg, S., 1972, *Gravitation and Cosmology* (Joh Wiley and Sons, New York, 1972).

Weinberg, S., 1995, *The Quantum Theory of Fields Volume I* (Cambridge University Press, New York, 1995).

Weyl, H., 1950, *Space, Time, Matter* (Dover, New York, 1950).

Weyl, H., (Tr. S. Pollard et al), 1987, *The Continuum* (Dover Publications, New York, 1987).

INDEX

About the Author

Stephen Blaha is a well known Physicist and Man of Letters with interests in Science, Society and civilization, the Arts, and Technology. He had an Alfred P. Sloan Foundation scholarship in college. He received his Ph.D. in Physics from Rockefeller University. He has served on the faculties of several major universities. He was also a Member of the Technical Staff at Bell Laboratories, a manager at the Boston Globe Newspaper, a Director at Wang Laboratories, and President of Blaha Software Inc and of Janus Associates Inc. (NH).

Among other achievements he was a co-discoverer of the "r potential" for heavy quark binding developing the first (and still the only demonstrable) non-abelian gauge theory with an "r" potential; first suggested the existence of topological structures in superfluid He-3; first proposed Yang-Mills theories would appear in condensed matter phenomena with non-scalar order parameters; first developed a grammar-based formalism for quantum computers and applied it to elementary particle theories; first developed a new form of quantum field theory without divergences (thus solving a major 60 year old problem that enabled a unified theory of the Standard Model and Quantum Gravity without divergences to be developed); first developed a formulation of complex General Relativity based on analytic continuation from real space-time; first developed a generalized non-homogeneous Robertson-Walker metric that enabled a quantum theory of the Big Bang to be developed without singularities at $t = 0$; first generalized Cauchy's theorem and Gauss' theorem to complex, curved multi-dimensional spaces; received Honorable Mention in the Gravity Research Foundation Essay Competition in 1978; first developed a physically acceptable theory of faster-than-light particles; first showed a universe with three complex spatial dimensions is icosahedral; first derived a composition of extrema method in the Calculus of Variations; first quantitatively suggested that inflationary periods in the history of the universe were not needed; first proved Gödel's Theorem implies Nature must be quantum; provided a new alternative to the Higgs Mechanism, and Higgs particles, to generate masses; first showed how to resolve logical paradoxes including Gödel's Undecidability Theorem by developing Operator Logic and Quantum Operator Logic; first developed a quantitative harmonic oscillator-like model of the life cycle, and interactions, of civilizations; first showed how equations describing superorganisms also apply to civilizations. A recent book shows his theory applies successfully to the past 14 years of history and to *new* archaeological data on Andean and Mayan civilizations as well as Early Anatolian and Egyptian civilizations.

He first developed an axiomatic derivation of the forms of The Standard Model with WIMPs from geometry – space-time properties – The faster than light Standard Model.

He has had a major impact on a succession of elementary particle theories: his Ph.D. thesis (1970), and papers, showed that quantum field theory calculations to all orders in ladder approximations could not give scaling deep inelastic electron-nucleon scattering. He later showed the eigenvalue equation for the fine structure constant α in Johnson-Baker-Willey QED had a zero at $\alpha = 1$ not 1/137 by solving the Schwinger-Dyson equations to all orders in an approximation that agreed with exact results to 8th order in α thus ending interest in this theory. In 1979 at Prof. Ken Johnson's (MIT) suggestion he calculated the proton-neutron mass difference in the MIT bag model and found the result had the wrong sign reducing interest in the bag model. These results all appear in Physical Review papers. In the 2000's he repeatedly pointed out the shortcomings of SuperString theory and showed that The Standard Model's form could be derived from space-time geometry by an extension of Lorentz transformations to faster than light transformations. This deeper space-time basis greatly increases the possibility that it is part of THE fundamental theory.

In graduate school (1965-71) he wrote substantial papers in elementary particles and group theory: The Inelastic E- P Structure Functions in a Gluon Model. Phys.Lett. B40:501-502,1972; Deep-Inelastic E-P Structure Functions In A Ladder Model With Spin 1/2 Nucleons, Phys.Rev. D3:510-523,1971; Continuum Contributions To The Pion Radius, Phys.Rev. 178:2167-2169,1969; Character Analysis of U(N) and SU(N), J.

Math. Phys. 10, 2156 (1969); and The Calculation of the Irreducible Characters of the Symmetric Group in Terms of the Compound Characters, (Published as Blaha's Lemma in D. E. Knuth's book: *The Art of Computer Programming Vols. 1 – 4*).

In the early 1980's Blaha was also a pioneer in the development of UNIX for financial, scientific and Internet applications: benchmarked UNIX versions showing that block size was critical for UNIX performance, developing financial modeling software, starting database benchmarking comparison studies, developing Internet-like UNIX networking (1982) and developing a hybrid shell programming technique (1982) that was a precursor to the PERL programming language. He was also the manager of the AT&T ten-year future products development database. His work helped lead to commercial UNIX on computers such as Sun Micros, IBM AIX minis, and Apple computers.

In the 1980's he pioneered the development of PC Desktop Publishing on laser printers. and was nominated for three "Awards for Technical Excellence" in 1987 by PC Magazine for PC software products that he designed and developed.

Recently he has developed a theory of Megaverses – actual universes of which our universe is one – with quantum particle-like properties based on the Wheeler-DeWitt equation of Quantum Gravity. He has developed a theory of a baryonic force, which had been conjectured many years ago, and estimated the strength of the force based on discrepancies in measurements of the gravitational constant G. This force, operative in 15-dimensinal space, can be used to escape from our universe in "uniships" which are the equivalent of the faster-than-light starships proposed in the author's earlier books. Thus travel to other universes, as well as to other stars is possible.

Blaha also considered the complexified Wheeler-DeWitt equation and showed that its limitation to real-valued coordinates and metrics generated a Cosmological Constant in the Einstein equations.

The author has also recently written a series of books on the serious problems of the United States and their solution as well as a book on the decline of Mankind that will follow from current social and genetic trends in Mankind.

In the past twelve years Dr. Blaha has written over 40 books on a wide range of topics. Some recent major works are: *From Asynchronous Logic to The Standard Model to Superflight to the Stars*, *All the Universe!, SuperCivilizations: Civilizations as Superorganisms, America's Future: an Islamic Surge, ISIS, al Qaeda, World Epidemics, Ukraine, Russia-China Pact, US Leadership Crisis,The Rises and Falls of Man – Destiny – 3000 AD: New Support for a Superorganism MACRO-THEORY of CIVILIZATIONS From CURRENT WORLD TRENDS and NEW Peruvian, Pre-Mayan, Mayan, Anatolian, and Early Egyptian Data, with a Projection to 3000 AD,* and *Mankind in Decline: Genetic Disasters, Human-Animal Hybrids, Overpopulation, Pollution, Global Warming, Food and Water Shortages, Desertification, Poverty, Rising Violence, Genocide, Epidemics, Wars, Leadership Failure.*

He has taught approximately 4,000 students in undergraduate, graduate, and postgraduate corporate education courses primarily in major universities, and large companies and government agencies.

The above paragraphs summarize much of his work over the past forty seven years. This work is fully documented. He continues to engage in research and writing at Blaha Research.

www.ingramcontent.com/pod-product-compliance
Lightning Source LLC
Chambersburg PA
CBHW082008190326
41458CB00010B/3122
* 9 7 8 0 9 8 9 3 8 2 6 8 7 *